T0184119

Lecture Notes in Computer Science   12603

More information about this subseries at http://www.springer.com/series/7412

Vincent Andrearczyk · Valentin Oreiller ·
Adrien Depeursinge (Eds.)

Head and Neck Tumor Segmentation

First Challenge, HECKTOR 2020
Held in Conjunction with MICCAI 2020
Lima, Peru, October 4, 2020
Proceedings

 Springer

Editors
Vincent Andrearczyk (iD)
HES-SO Valais-Wallis University
of Applied Sciences and Arts
Western Switzerland
Sierre, Switzerland

Valentin Oreiller (iD)
HES-SO Valais-Wallis University
of Applied Sciences and Arts
Western Switzerland
Sierre, Switzerland

Adrien Depeursinge (iD)
HES-SO Valais-Wallis University
of Applied Sciences and Arts
Western Switzerland
Sierre, Switzerland

ISSN 0302-9743   ISSN 1611-3349   (electronic)
Lecture Notes in Computer Science
ISBN 978-3-030-67193-8   ISBN 978-3-030-67194-5   (eBook)
https://doi.org/10.1007/978-3-030-67194-5

LNCS Sublibrary: SL6 – Image Processing, Computer Vision, Pattern Recognition, and Graphics

This Springer imprint is published by the registered company Springer Nature Switzerland AG
The registered company address is: Gewerbestrasse 11, 6330 Cham, Switzerland

Preface

Radiomics, i.e. high-throughput biomedical image analysis, has gained particular interest in the last decade to improve personalized oncology. It relies on the extraction of visual features contained in a volume of interest, i.e. the tumor region. One of the main hindrances to the development of this field is the lack of large annotated cohorts including tumor and metastasis contours. This is mainly due to the difficulty, time and ultimately cost of obtaining manually segmented images. Therefore, the automatic segmentation of biomedical images is one of the first problems that must be solved to generate big data allowing the development of more precise and fully automatic radiomics models for personalized medicine.

Head and Neck (H&N) cancer is still among the deadliest cancers. Choosing and tailoring treatment for a given patient would highly benefit from an automatic targeted quantitative assessment via radiomics models. Although required for large-scale radiomics studies, H&N tumor segmentation in fluorodeoxyglucose (FDG)-Positron Emission Tomography (PET)/Computed Tomography (CT) scans remains poorly studied. For these reasons, we deemed it timely to organize a challenge on this topic. In order to evaluate and compare the current state-of-the-art methods for automatic H&N tumor segmentation, we proposed the HEad and neCK TumOR (HECKTOR 2020)[1] segmentation challenge hosted by the International Conference on Medical Image Computing and Computer Assisted Intervention (MICCAI 2020)[2]. In the context of this challenge, a dataset of 204 delineated PET/CT images was made available for training as well as 53 PET/CT images for testing. Various deep learning methods were developed by the participants with excellent results. The submitted automatic segmentations were evaluated using standard evaluation metrics (Dice score coefficient) using annotations of the test set that were not provided to the participants. These results were presented at the half-day event on October 4, 2020 at a designated satellite event of the MICCAI 2020. Following this event the leaderboard remains open for new submissions.

Among the 18 teams who submitted their segmentation outcome, 10 teams submitted a paper describing their method and results. All the submitted papers were accepted after a single-blind review process with a minimum of two reviewers per paper. The present volume gathers these participants' papers as well as our overview paper (which also underwent the same reviewing process as the participants' papers).

---

[1] www.aicrowd.com/challenges/hecktor.

[2] www.miccai2020.org, as of November 2020.

We thank the committee members, the participants, the MICCAI organizers, the reviewers, and our sponsor Siemens Healthineers Switzerland.

December 2020

Vincent Andrearczyk
Valentin Oreiller
Adrien Depeursinge

Organization

General Chairs

Vincent Andrearczyk   University of Applied Sciences and Arts Western Switzerland, Switzerland

Valentin Oreiller   University of Applied Sciences and Arts Western Switzerland, Switzerland

Adrien Depeursinge   University of Applied Sciences and Arts Western Switzerland, Switzerland

Program Committee Chairs

Mario Jreige   Lausanne University Hospital, Switzerland
Martin Vallières   University of Sherbrooke, Canada
Joel Castelli   University of Rennes 1, France
Hesham Elhalawani   Cleveland Clinic Foundation, USA
Sarah Boughdad   Lausanne University Hospital, Switzerland
John O. Prior   Lausanne University Hospital, Switzerland

Additional Reviewers

Yashin Dicente
Pierre Fontaine
Alba Garcia
Mara Graziani
Oscar Jimenez
Niccolò Marini
Henning Müller
Sebastian Otálora
Annika Reinke

Contents

Overview of the HECKTOR Challenge at MICCAI 2020: Automatic Head and Neck Tumor Segmentation in PET/CT

Vincent Andrearczyk[1]([✉]), Valentin Oreiller[1,2], Mario Jreige[2],
Martin Vallières[3], Joel Castelli[4,5,6], Hesham Elhalawani[7], Sarah Boughdad[2],
John O. Prior[2], and Adrien Depeursinge[1,2]

[1] Institute of Information Systems, School of Management, HES-SO Valais-Wallis
University of Applied Sciences and Arts Western Switzerland, Sierre, Switzerland
vincent.andrearczyk@hevs.ch
[2] Centre Hospitalier Universitaire Vaudois (CHUV), Lausanne, Switzerland
[3] Department of Computer Science, University of Sherbrooke,
Sherbrooke, QC, Canada
[4] Radiotherapy Department, Cancer Institute Eugène Marquis, Rennes, France
[5] INSERM, U1099, Rennes, France
[6] University of Rennes 1, LTSI, Rennes, France
[7] Department of Radiation Oncology, Cleveland Clinic Foundation,
Cleveland, OH, USA

Abstract. This paper presents an overview of the first HEad and neCK TumOR (HECKTOR) challenge, organized as a satellite event of the 23rd International Conference on Medical Image Computing and Computer Assisted Intervention (MICCAI) 2020. The task of the challenge is the automatic segmentation of head and neck primary Gross Tumor Volume in FDG-PET/CT images, focusing on the oropharynx region. The data were collected from five centers for a total of 254 images, split into 201 training and 53 testing cases. The interest in the task was shown by the important participation with 64 teams registered and 18 team submissions. The best method obtained a Dice Similarity Coefficient (DSC) of 0.7591, showing a large improvement over our proposed baseline method with a DSC of 0.6610 as well as inter-observer DSC agreement reported in the literature (0.69).

Keywords: Automatic segmentation · Challenge · Medical imaging · Head and neck cancer · Oropharynx

1 Introduction: Research Context

The prediction of disease characteristics using quantitative image biomarkers from medical images (i.e. radiomics) has shown tremendous potential to optimize

---

V. Andrearczyk and V. Oreiller—Equal contribution.

V. Andrearczyk et al. (Eds.): HECKTOR 2020, LNCS 12603, pp. 1–21, 2021.
https://doi.org/10.1007/978-3-030-67194-5_1

patient care, particularly in the context of Head and Neck (H&N) tumors [20]. FluoroDeoxyGlucose (FDG)-Positron Emission Tomography (PET) and Computed Tomography (CT) imaging are the modalities of choice for the initial staging and follow-up of H&N cancer. Yet, radiomics analyses rely on an expensive and error-prone manual annotation process of Volumes of Interest (VOI) in three dimensions. The automatic segmentation of H&N tumors from FDG-PET/CT images could therefore enable the validation of radiomics models on very large cohorts and with optimal reproducibility. Besides, automatic segmentation algorithms could enable a faster clinical workflow. By focusing on metabolic and morphological tissue properties respectively, PET and CT modalities include complementary and synergistic information for cancerous lesion segmentation. The HEad and neCK TumOR (HECKTOR)[1] challenge aims at identifying the best methods to leverage the rich bi-modal information in the context of H&N primary tumor segmentation. This precious knowledge will be transferable to many other cancer types where PET/CT imaging is relevant, enabling large-scale and reproducible radiomics studies.

The potential of PET information for automatically segmenting tumors has been long exploited in the literature. For an in-depth review of automatic segmentation of PET images in the pre-deep learning era, see [5] covering methods such as thresholding, active contours and mixture models. The first challenge on tumor segmentation in PET images was proposed at MICCAI 2016[2] by Hatt et al. [8]. The need for a standardized evaluation of PET automatic segmentation methods and a comparison study between all the current algorithms was highlighted in [9]. Multi-modal analyses of PET and CT images have also recently been proposed for different tasks, including lung cancer segmentation in [11,12,25,26] and bone lesion detection in [22]. In [2], we developed a baseline Convolutional Neural Network (CNN) approach based on a leave-one-center-out cross-validation on the training data of the HECKTOR challenge. Promising results were obtained with limitations that motivated additional data curation, data cleaning and the creation of this challenge. This challenge builds upon these works by comparing, on a publicly available dataset, recent segmentation architectures as well as the complementarity of the two modalities on a task of primary Gross Tumor Volume (GTVt) segmentation of H&N tumor in the oropharynx region. The proposed dataset comprises data from five centers. Four centers are used for the training data and one for testing. The task is challenging due to, among others, the variation in image acquisition and quality across centers (test set from an unseen center) and the presence of lymph nodes with high metabolic responses in the PET images.

The critical consequences of the lack of quality control in challenge designs were shown in [14], including reproducibility and interpretation of the results often hampered by the lack of provided relevant information, and non-robust ranking of algorithms. Solutions were proposed in the form of the Biomedical

---

[1] www.aicrowd.com/challenges/hecktor, as of October 2020.

[2] https://portal.fli-iam.irisa.fr/petseg-challenge/overview#_ftn1, as of October 2020.

Image Analysis challengeS (BIAS) [15] guidelines for reporting the results. This paper presents an overview of the challenge following these guidelines.

Individual participants' papers were submitted to the challenge organizers, reporting methods and results. Reviews were organized by the organizers and the papers of the participants are published in the LNCS challenges proceedings [4, 6,10,13,17,18,21,23,24,27].

The paper is organized as follows. The challenge design and data description are described in Sect. 2. The main results of the challenge are reported in Sect. 3 and discussed in Sect. 4. Finally, Sect. 5 concludes this paper.

2 Methods: Reporting of Challenge Design

A summary of the information on the challenge organization is provided in Appendix 1, following the BIAS recommendations.

2.1 Mission of the Challenge

Biomedical Application
The participating algorithms target the following fields of application: diagnosis, prognosis and research. The participating teams' algorithms were designed for image segmentation, more precisely, classifying voxels as either tumor or background.

Cohorts
As suggested in [15], we refer to the patients from whom the image data were acquired as the cohort. The target cohort[3] comprises patients received for initial staging of H&N cancer. The clinical goals are two-fold; the automatically segmented regions can be used as a basis for (i) treatment planning in radiotherapy, (ii) further radiomics studies to predict clinical outcomes such as overall patient survival, disease-free survival, tumor aggressivity. In the former case (i), the regions will need to be further refined or extended for optimal dose delivery and control. The challenge cohort[4] includes patients with histologically proven H&N cancer who underwent radiotherapy treatment planning. The data were acquired from five centers (four for the training and one for the testing) with variation in the scanner manufacturers and acquisition protocols. The data include PET and CT imaging modalities as well as patient information including age, sex and acquisition center. A detailed description of the annotations is provided in Sect. 2.2.

---

[3] The target cohort refers to the subjects from whom the data would be acquired in the final biomedical application. It is mentioned for additional information as suggested in BIAS, although all data provided for the challenge are part of the challenge cohort.

[4] The challenge cohort refers to the subjects from whom the challenge data were acquired.

Target Entity
The data origin, i.e. the region from which the image data were acquired, varied from the head region only to the whole body. While we provided the data as acquired, we limited the analysis to the oropharynx region and provided an automatically detected bounding box locating the oropharynx region [1], as illustrated in Fig. 1.

Assessment Aim
The assessment aim is the following; evaluate the feasibility of fully automatic GTVt segmentation for H&N cancers in the oropharyngeal region via the identification of the most accurate segmentation algorithm. The performance of the latter is identified by computing the Dice Similarity Coefficient (DSC) between prediction and manual expert annotations. The individual DSC scores are averaged for all test patients and the ranking is based on this average score. DSC measures volumetric overlap between segmentation results and annotations. It is a good measure of segmentation for imbalanced segmentation problems, i.e. the region to segment is small as compared to the image size. DSC is commonly used in the evaluation and ranking of segmentation algorithms and particularly tumor segmentation tasks [7,16].

Missing values (i.e. missing predictions on one or multiple patients), did not occur in the submitted results, but will be treated as DSC of zero if it occurs in future submissions on the open leaderboard. In case of tied rank, very unlikely due to the computation of the results (average of 53 DSCs), we will consider the precision as the second ranking metric.

A statistical analysis is performed to statistically compare the performance of the algorithms using a Wilcoxon-signed rank test.

2.2   Challenge Dataset

Data Source
The data were acquired from five centers as listed in Table 1. It consists of PET/CT images of patients with H&N cancer located in the oropharynx region. The scanners (devices) and imaging protocols used to acquire the data are described in Table 2. Additional information about the image acquisition is provided in Appendix 2.

Table 1. List of the hospital centers in Canada (CA) and Switzerland (CH) and number of cases, with a total of 201 training and 53 test cases.

Center	Split	# cases
HGJ: Hôpital Général Juif, Montréal, CA	Train	55
CHUS: Centre Hospitalier Universitaire de Sherbooke, Sherbrooke, CA	Train	72
HMR: Hôpital Maisonneuve-Rosemont, Montréal, CA	Train	18
CHUM: Centre Hospitalier de l'Université de Montréal, Montréal, CA	Train	56
Total	Train	201
CHUV: Centre Hospitalier Universitaire Vaudois, CH	Test	53

Table 2. List of scanners used in the different centers.

Center	Device
HGJ	hybrid PET/CT scanner (Discovery ST, GE Healthcare)
CHUS	hybrid PET/CT scanner (GeminiGXL 16, Philips)
HMR	hybrid PET/CT scanner (Discovery STE, GE Healthcare)
CHUM	hybrid PET/CT scanner (Discovery STE, GE Healthcare)
CHUV	hybrid PET/CT scanner (Discovery D690 TOF, GE Healthcare)

Training and Test Case Characteristics
The training data comprise 201 cases from four centers (HGJ, HMR[5], CHUM and CHUS). Originally, the dataset in [20] contained 298 cases, among which we selected the cases with oropharynx cancer. The test data comprise 53 cases from another fifth center (CHUV). Examples of PET/CT images of each center are shown in Fig. 1. Each case comprises a CT image, a PET image and a GTVt mask (for the training cases) in the Neuroimaging Informatics Technology Initiative (NIfTI) format, as well as patient information (age, sex) and center. A bounding box locating the oropharynx region was also provided (details of the automatic region detection can be found in [1]).

Finally, to provide a fair comparison, participants who wanted to use additional external data for training were asked to also report results using only the HECKTOR data and discuss differences in the results.

Annotation Characteristics
Initial annotations, i.e. 3D contours of the GTVt, were made by expert radiation oncologists and were later modified by a VOI quality control and correction as described later. Details of the initial annotations of the training set can be found in [20]. In particular, 40% (80 cases) of the training radiotherapy contours were directly drawn on the CT of the PET/CT scan and thereafter used for treatment planning. The remaining 60% of the training radiotherapy contours were drawn on a different CT scan dedicated to treatment planning and were then registered to the FDG-PET/CT scan reference frame using intensity-based free-form deformable registration with the software MIM (MIM software Inc., Cleveland, OH). The initial contours of the test set were all directly drawn on the CT of the PET/CT scan.

VOI quality control and correction were supervised by an expert who is both radiologist and nuclear medicine physician. Two non-experts (organizers of the challenge) made an initial cleaning in order to facilitate the expert's work. The expert either validated or edited the VOIs. The Siemens Syngo.Via RT Image Suite was used to edit the contours in 3D with fused PET/CT images. The main points corrected during the data curation are listed in the following.

- All annotations were originally performed in a radiotherapy context. They were potentially inadequate for radiomics studies as too large, often including

---

[5] For simplicity, these centers were renamed CHGJ and CHMR during the challenge.

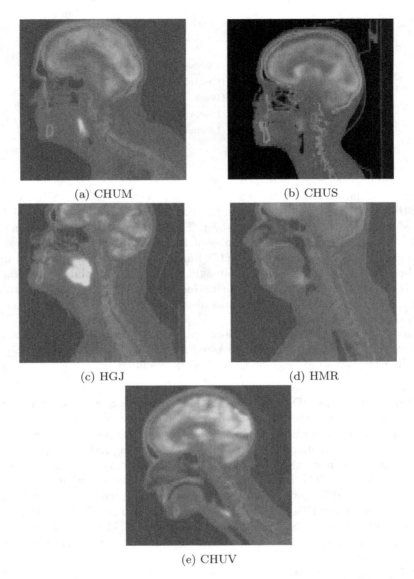

(a) CHUM

(b) CHUS

(c) HGJ

(d) HMR

(e) CHUV

Fig. 1. Case examples of 2D sagittal slices of fused PET/CT images from each of the five centers.

air in the trachea and various tissues surrounding the tumor. The primary tumors were delineated as close as possible to the real tumoral volume.
- Some annotations were originally drawn on a distinct CT scan dedicated to treatment planning. In this case, the contours were registered to the PET/CT scans (more details in [20]). Some registrations failed and had to be corrected.

- Some VOIs included both the primary tumor (GTVt) and lymph nodes (GTVn) without distinction. Separating the primary tumor from the lymph nodes was essential as they carry different information and should not be grouped.
- One contour (HGJ069) was flipped symmetrically on the A/P plane.
- Some annotations were missing and had to be drawn from scratch by the expert.

Data Preprocessing Methods
No preprocessing was performed on the images to reflect the diversity of clinical data and to leave full flexibility to the participants. However, we provided various pieces of code to load, crop, resample the data, train a baseline CNN (NiftyNet) and evaluate the results on our GitHub repository[6]. This code was provided as a suggestion to help the participants and to maximize transparency, but the participants were free to use other methods.

Sources of Errors
According to Gudi et al. [7], in the context of radiotherapy planning, one can expect an inter-observer DSC in tumor segmentation of 0.57 and 0.69 on CT and PET/CT respectively, highlighting the difficulty of the task. A source of error therefore originates from the degree of subjectivity in the annotation and correction of the expert. For most patients, the tumors were contoured on another CT scan, then the two CTs were registered and the annotations were transformed according to the registrations. Thus, a major source of error came from this registration step.

Another source of error comes from the lack of CT images with a contrast agent for a more accurate delineation of the primary tumor.

Institutional Review Boards
Institutional Review Boards (IRB) of all participating institutions permitted the use of images and clinical data, either fully anonymized or coded, from all cases for research purposes, only. Retrospective analyses were performed following the relevant guidelines and regulations as approved by the respective institutional ethical committees with protocol numbers: MM-JGH-CR15-50 (HGJ, CHUS, HMR, CHUM) and CER-VD 2018-01513 (CHUV).

2.3 Assessment Method

Participants were given access to the test cases without the ground truth annotations and were asked to submit the results of their algorithms on the test cases on the AIcrowd platform.

Results were ranked using the (3D) Dice Similarity Coefficient (DSC) computed on images cropped using the provided bounding boxes (see Sect. 2.2) in the original CT resolution as:

$$DSC = \frac{2TP}{2TP + FP + FN},\qquad(1)$$

---

[6] github.com/voreille/hecktor, as of October 2020.

where TP, FP and FN are the number of true positive, false positive and false negative pixels, respectively. If the submitted results were in a resolution different from the CT resolution, we applied nearest-neighbor interpolation before evaluation. We also computed other metrics for comparison, namely precision ($\frac{TP}{TP+FP}$) and recall ($\frac{TP}{TP+FN}$) to investigate whether the method was rather providing a large FP or FN rate. The evaluation implementation can be found on our GitHub repository[7] and was provided to maximize transparency.

Each participating team had the opportunity to submit up to five (valid) runs. The best result of each team was used in the final ranking, which is detailed in Sect. 3 and discussed in Sect. 4.

3   Results: Reporting of Challenge Outcome

3.1   Participation

We received and approved, as of Sept. 10 2020 (submission deadline), 85 signed end-user-agreements. At the same date, the number of registered teams was 64. A team is made of at least one participant and not all participants that signed the end-user-agreement registered a team. Each team could submit up to five valid submissions. By the submission deadline, we had received 83 results submissions, including valid and invalid ones (i.e. non graded due to format errors). For the first iteration of the challenge, these numbers are high and show an important interest in the task.

3.2   Algorithms Summary

Organizers' Baselines
We trained several baseline models using standard 3D and 2D U-Nets [19] as in our preliminary results in [2] (the data were different). We trained on multi-modal PET/CT as well as individual modalities with a non-weighted Dice (i.e. based on DSC) and cross-entropy loss and without data augmentation.

Participants' Methods
In [10], Iantsen et al. proposed a model based on a U-Net architecture with residual layers and supplemented with 'Squeeze and Excitation' (SE) normalization, previously developed by the same authors for brain tumor segmentation. An unweighted sum of soft Dice loss and Focal Loss was used for training. The test results were obtained as an ensemble of eight models trained and validated on different splits of the training set. No data augmentation was performed.

In [13], Ma and Yang used a combination of U-Nets and hybrid active contours. First, 3D U-Nets are trained to segment the tumor (with a cross-validation on the training set). Then, the segmentation uncertainty is estimated by model ensembles on the test set to select the cases with high uncertainties. Finally, the authors used a hybrid active contour model to refine the high uncertainty cases.

---

[7] github.com/voreille/hecktor/tree/master/src/evaluation, as of October 2020.

The U-Nets were trained with an unweighted combination of Dice loss and top-K loss. No data augmentation was used.

In [27], Zhu et al. used a two steps approach. First, a classification network (based on ResNet) selects the axial slices which may contain the tumor. These slices are then segmented using a 2D U-Net to generate the binary output masks. Data augmentation was applied by shifting the crop around the provided bounding boxes and the U-Net was trained with a soft Dice loss. The preprocessing includes clipping the CT and the PET, standardizing the HU within the cropped volume and scaling the range of the PET to correspond to the CT range by dividing it by a factor of 10.

In [24], Yuan proposed to integrate information across different scales by using a dynamic Scale Attention Network (SA-Net), based on a U-Net architecture. Their network incorporates low-level details with high-level semantics from feature maps at different scales. The network was trained with standard data augmentation and with a Jaccard distance loss, previously developed by the authors. The results on the test set were obtained as an ensemble of ten models.

In [4], Chen et al. proposed a three-step framework with iterative refinement of the results. In this approach, multiple 3D U-Nets are trained one-by-one using a Dice loss without data augmentation. The predictions and features of previous models are captured as additional information for the next one to further refine the segmentation.

In [6], Ghimire et al. developed a patch-based approach to tackle the memory issue associated with 3D images and networks. They used an ensemble of conventional convolutions (with small receptive fields capturing fine details) and dilated convolutions (with a larger receptive field of capturing global information). They trained their model with a weighted cross-entropy and dice loss and random left-right flips of the patches were applied for data augmentation. Finally, an ensemble of the best two models selected during cross-validation was used for predicting the segmentation of the test data.

In [23], Yousefirizi and Rahmim proposed a deep 3D model based on SegAN, a generative adversarial network (GAN) for medical image segmentation. An improved polyphase V-net (to help preserve boundary details) is used for the generator and the discriminator network has a similar structure to the encoder part of the former. The networks were trained using a combination of Mumford-Shah (MS) and multi-scale Mean Absolute Error (MAE) losses, without data augmentation.

In [21], Xie and Peng proposed a 3D scSE nnU-Net model, improving upon the 3D nnU-Net by integrating the spatial and channel 'Squeeze and Excitation' (scSE) blocks. They trained the model with a weighted combination of Dice and cross-entropy losses, together with standard data augmentation techniques (rotation, scaling etc.). To preprocess the CT images an automated level-window-like clipping of intensity values is performed based on the 0.5 and 99.5th percentile of these values. The intensity values of the PET are standardized by subtracting the mean and then, by dividing by the standard deviation of the image.

In [17], Naser et al. used a variant of 2D and 3D U-Net (we report the best result, with the 3D model). The models were trained with a combination of Dice and cross-entropy losses with standard data augmentation.

In [18], Rao et al. proposed an ensemble of two methods, namely a 3D U-Net and another 2D U-Net variant with 3D context. A top-k loss was used to train the models without data augmentation.

In Table 3, we summarize some of the main components of the participants' algorithms, including model architecture, preprocessing, training scheme and postprocessing.

Table 3. Summary of the algorithms with some of the main components: 2D or 3D U-Net, resampling, preprocessing, training or testing data augmentation, loss used for optimization, ensemble of multiple models for test prediction and postprocessing of the results. We use the following abbreviations for the preprocessing: clipping (C), standardization (S), and if it is applied only to one modality, it is specified in parentheses. For the image resampling, we specify whether the algorithms use isotropic (I) or anisotropic (A) resampling and nearest neighbor (NN), linear (L) or cubic (Cu) interpolation. We use the following abbreviation for the loss: Cross-Entropy (CE), Mumford-Shah (MS) and Mean Absolute Error (MAE). More details can be found in the respective participants' publications.

Team	2D/3D	preproc.	resampling	augm.	loss	ensemble	postproc.
andrei.iantsen [10]	3D	C+S	I/L	✓	soft Dice+Focal	✓	✗
junma [13]	3D	S(PET)	I/Cu	✗	Dice+Top-K	✓	✓
badger [21]	3D	C(CT)+S(PET)	A/Cu	✓	Dice+CE	✗	✗
deepX [24]	3D	C(CT)+S	I/L	✓	Jaccard distance	✓	✗
AIView_sjtu [4]	3D	C+S	A/NN	✓	Dice	✗	✗
xuefeng [6]	3D	C(CT)+S	A/L	✓	Dice+CE	✓	✓
QuritLab [23]	3D	S	I/L	✗	MS+MAE	✗	✗
HFHSegTeam [27]	2D	C+S	I/L	✓	soft Dice	✗	✗
Fuller_MDA_Lab [17]	3D	C+S	A/Cu	✓	Dice+CE	✗	✗
Maastro-Deep-Learning [18]	2D/3D	C	A/Cu	✗	Top-K	✓	✓
Our baseline 3D PET/CT	3D	C+S	I/Cu	✗	Dice+CE	✗	✗
Our baseline 2D PET/CT	2D	C+S	I/Cu	✗	Dice+CE	✗	✗

3.3   Results

The results, including average DSC, precision, recall and challenge rank are summarized in Table 4. Our baseline method, developed in [2] and provided to participants as an example on our GitHub repository, obtains an average DSC of 0.6588 and 0.6610 with the 2D and 3D implementations respectively. Results on individual modalities are also reported for comparison. The results from the participants range from an average DSC of 0.5606 to 0.7591. Iantsen et al. [10] (participant *andrei.iantsen*) obtained the best overall results with an average DSC of 0.7591, an average precision of 0.8332 and an average recall of 0.7400. These results (DSCs) are not significantly higher than the second best participant [13] (p-value 0.3501 with a one-tail Wilcoxon signed-rank test) and

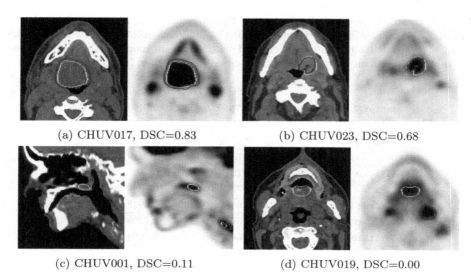

(a) CHUV017, DSC=0.83   (b) CHUV023, DSC=0.68

(c) CHUV001, DSC=0.11   (d) CHUV019, DSC=0.00

Fig. 2. Examples of results of the winning team (andrei.iantsen [10]). The automatic segmentation results (green) and ground truth annotations (red) are displayed on an overlay of 2D slices of PET (right) and CT (left) images. The reported DSC is computed on the whole image (see Eq. 1). (a), (b) Excellent segmentation results, detecting the GTVt of the primary oropharyngeal tumor localized at the base of the tongue and discarding the laterocervical lymph nodes despite high FDG uptake on PET. (c) Incorrect segmentation of the top volume at the level of the soft palate; (d) Incorrect segmentation of the tongue due to an abnormal FDG uptake (possible reasons include prior surgical intervention of the tongue, chewing gum and involuntary movements of the tongue). (Color figure online)

are significantly higher than the third best participant (p-value 0.0041 with the same test). Across all participants, the average precision ranges from 0.5850 to 0.8479. The recall ranges from 0.5022 to 0.8534, with the latter surprisingly obtained by the 3D PET/CT baseline (although with low precision, reflecting an over-segmentation as compared to other algorithms' outputs). Note that two participants decided to withdraw their submissions due to very low scores. We allowed them to do so since their low scores were due to incorrect postprocessing (e.g. setting incorrect pixel spacing, or image origin) and were not representative of the performance of their algorithms.

Examples of segmentation results (true positives on top row, and false positives on bottom row) are shown in Fig. 2.

Table 4. Summary of the challenge results. The average DSC, precision and recall are reported for the baseline algorithms and for the different teams (best result of each team). The participant names are reported when no team name was provided. The ranking is only provided for teams that presented their method in a paper submission.

Team	DSC	Precision	Recall	Rank
andrei.iantsen [10]	**0.7591**	0.8332	0.7400	1
junma [13]	0.7525	0.8384	0.7471	2
badger [21]	0.7355	0.8326	0.7023	3
deepX [24]	0.7318	0.7851	0.7319	4
AIView_sjtu [4]	0.7241	**0.8479**	0.6701	5
DCPT	0.7049	0.7651	0.7047	-
xuefeng [6]	0.6911	0.7525	0.6928	6
ucl_charp	0.6765	0.7231	0.7256	-
QuritLab [23]	0.6677	0.7290	0.7164	7
Unipa	0.6674	0.7143	0.7039	-
Our baseline 3D PET/CT	0.6610	0.5909	**0.8534**	-
Our baseline 2D PET/CT	0.6588	0.6241	0.7629	-
HFHSegTeam [27]	0.6441	0.6938	0.7014	8
UESTC_501	0.6381	0.6455	0.6874	-
Fuller_MDA_Lab [17]	0.6373	0.7546	0.6283	9
Our baseline 3D PET	0.6306	0.5768	0.8214	-
Our baseline 2D PET	0.6284	0.6470	0.6666	-
Maastro-Deep-Learning [18]	0.5874	0.6560	0.6141	10
Yone	0.5737	0.6606	0.5590	-
SC_109	0.5633	0.7652	0.5022	-
Roque	0.5606	0.5850	0.6843	-
Our baseline 2D CT	0.3071	0.3477	0.3574	-
Our baseline 3D CT	0.2729	0.2154	0.5874	-

4   Discussion: Putting the Results into Context

4.1   Outcome and Findings

A major benefit of this challenge is to compare various algorithms developed by teams from all around the world on the same dataset and task, with held-out test data.

We distinguish here between the technical and biomedical impact. The main technical impact of the challenge is the comparison of state-of-the-art algorithms on the provided data. We identified key elements for addressing the task: 3D U-Net, preprocessing, normalization, data augmentation and ensembling, as summarized in Table 3. The main biomedical impact of the results is the opportunity to generate large cohorts with automatic tumor segmentation for comprehensive radiomics studies.

The best methods obtain excellent results with DSCs above 0.75, better than[8] reported inter-observer variability (DSCs of 0.57 and 0.69 on CT and PET-CT respectively) [7]. Note that without injected contrast CT, delineating the exact contour of the tumor is very difficult. Thus, the inter-observer DSC could be low only due to disagreements at the border of the tumor, without taking into account the error rate due to the segmentation of non-malignant structures (if any). For that reason, defining the task as solved solely based on the DSC is not sufficient. In the context of this challenge, we can therefore define the task as solved if the algorithms follow these three criteria:

1. Higher or similar DSC than inter-observers agreement.
2. Detect all the primary tumors in the oropharynx region (i.e. segmentation not evaluated at the pixel level, rather at the occurrence level).
3. Similarly, detect only the primary tumors in the oropharynx region (discarding lymph nodes and other potentially false positives).

According to these criteria, the task is partially solved. The first criterion, evaluating the segmentation at the pixel level, is fulfilled. At the occurrence level (criteria 2 and 3), however, even the algorithms with the highest DSC output FP and FN regions. These errors are generally made in very difficult cases and we should further evaluate their source, e.g. Fig. 2c and 2d. Besides, there is still a lot of work to do on highly related tasks, including the segmentation of lymph nodes, the development of super-annotator ground truth as well as the agreement of multiple annotators, and, finally, the prediction of patient outcome following the tumor segmentation.

Following the analysis of poorly segmented cases, we identified several key elements that cause the algorithms to fail. These elements are as follows; low FDG uptake on PET, primary tumor that looks like a lymph node, abnormal uptake in the tongue and tumor present at the border of the oropharynx region. Some examples are illustrated in Fig. 1. Understanding these errors will lead to better methods and to a more targeted task for the next iteration of this challenge.

4.2   Limitations of the Challenge

The dataset provided in this challenge suffers from several limitations. First, the contours were mainly drawn based on the PET/CT fusion which is not sufficient to clearly delineate the tumor. Other methods such as MRI with gadolinium or contrast CT are the gold standard to obtain the true contours for radiation oncology. Since the target clinical application is radiomics, however, the precision of the contours is not as important as for radiotherapy planning.

---

[8] These values are reported only to give an idea of inter-observer variability on a similar task reported in the literature. The datasets are different and the comparison is limited. In future work, we will compute the inter-observer agreement on the challenge data.

Another limitation comes from the definition of the task, only one segmentation was drawn on the fusion of PET and CT. For radiomic analysis, it could be beneficial to consider one segmentation per modality since the PET signal is often not contained in the fusion-based segmentation due to the poor spatial resolution of this modality.

4.3   Lessons Learned and Next Steps

The guidelines, requirements and review process of the MICCAI submission helped us to design the challenge and to consider as much as possible the potential difficulties that could arise during the challenge.

Feedback from Participants
Relevant feedback was provided by the participants in the form of discussion and survey. They overall rated the quality of the data as good and the timing (see "Challenge schedule" in Appendix 1) appropriate. Participants were particularly interested in extending the challenge task to the segmentation of lymph nodes and, more moderately, in a radiomics task.

Future of the Challenge
The leaderboard remains open on the AIcrowd platform[9]. Participants can continue to develop new segmentation algorithms and compare their results with the existing ones.

Potentially, in the next edition (HECKTOR 2021), the participants will be asked to segment also the lymph nodes and/or to perform a radiomics study. We will also try to increase the size of the dataset with new training and test cases from other centers.

Finally, radiomics studies were proposed in [3, 20] to predict the prognosis of patients with H&N cancer in a non-invasive fashion. A limitation of these studies is that they were validated on 100 to 400 patients. Larger cohorts are required for estimating the generalization in real clinical settings. Manual annotations in 3D are tedious and error-prone, and the automatic tumor segmentation is an important step for large scale radiomics studies. To evaluate the feasibility of using automatic segmentation for radiomics studies, we will compare the automatic annotations to the manually delineated ones in a future work.

5   Conclusions

This paper presented a general overview of the HECKTOR challenge including the data, the participation, main results and discussions. The proposed task was the segmentation of the primary tumor in oropharyngeal cancer. The participation was relatively good with 18 results submissions and 10 participant's papers. This participation in the first edition of the HECKTOR challenge showed a high interest in automatic lesion segmentation for H&N cancer.

---

[9] www.aicrowd.com/challenges/miccai-2020-hecktor/leaderboards.

The task proposed this year was to segment the primary tumor in PET/CT images. This task is not as simple as thresholding the PET image since we target only the primary tumor and the region covered by high PET activation is often too large, going beyond the limits of the tumor tissues. Deep learning methods based on U-Net models were mostly used in the challenge. Interesting ideas were implemented to combine PET and CT complementary information. Model ensembling, as well as data preprocessing and augmentation, seem to have played an important role in achieving top-ranking results.

Acknowledgments. The organizers thank all the teams for their participation and valuable work. This challenge and the winner prize are sponsored by Siemens Healthineers Switzerland. This work was also partially supported by the Swiss National Science Foundation (SNSF, grant 205320_179069) and the Swiss Personalized Health Network (SPHN, via the IMAGINE and QA4IQI projects).

Appendix 1: Challenge Information

In this appendix, we list important information about the challenge as suggested in the BIAS guidelines [15].

Challenge Name
HEad and neCK TumOR segmentation challenge (HECKTOR) 2020.

Organizing Team
(Authors of this paper) Vincent Andrearczyk, Valentin Oreiller, Martin Vallières, Joel Castelli, Mario Jreige, John O. Prior and Adrien Depeursinge.

Life Cycle Type
A fixed submission deadline was set for the challenge results. Open online leaderboard following the conference.

Challenge Venue and Platform
The challenge is associated with MICCAI 2020. Information on the challenge is available on the website, together with the link to download the data, the submission platform and the leaderboard[10].

Participation Policies

(a) Algorithms producing fully-automatic segmentation of the test cases were allowed.
(b) The data used to train algorithms was not restricted. If using external data (private or public), participants were asked to also report results using only the HECKTOR data.
(c) Members of the organizers' institutes could participate in the challenge but were not eligible for awards.
(d) The award was 500 euros, sponsored by Siemens Healthineers Switzerland.

---

[10] www.aicrowd.com/challenges/hecktor.

(e) Policy for results announcement: The results were made available on the AIcrowd leaderboard and the best three results were announced publicly. Once participants submitted their results on the test set to the challenge organizers via the challenge website, they were considered fully vested in the challenge, so that their performance results (without identifying the participant unless permission is granted) became part of any presentations, publications, or subsequent analyses derived from the challenge at the discretion of the organizers.

(f) Publication policy: This overview paper was written by the organizing team's members. The participating teams were encouraged to submit a paper describing their method. The participants can publish their results separately elsewhere when citing the overview paper, and (if so) no embargo will be applied.

Submission Method

Submission instructions are available on the website[11] and are reported in the following. Results should be provided as a single binary mask (1 in the predicted GTVt) *.nii.gz* file per patient in the CT original resolution and cropped using the provided bounding boxes. The participants should pay attention to saving NIfTI volumes with the correct pixel spacing and origin with respect to the original reference frame. The .nii files should be named [PatientID].nii.gz, matching the patients' file names, e.g.. *CHUV001.nii.gz* and placed in a folder. This folder should be zipped before submission. If results were submitted without cropping and/or resampling, we employed nearest-neighbor interpolation given that the coordinate system is provided. Participants were allowed five valid submissions. The best result was reported for each team.

Challenge Schedule

The schedule of the challenge, including modifications, is reported in the following.

- the release date of the training cases: ~~June 01 2020~~ June 10 2020
- the release date of the test cases: Aug. 01 2020
- the results submission date(s): opens Sept. 01 2020 closes Sept. 10 2020
- paper submission deadline: ~~Sept. 18 2020~~ Sept. 15 2020
- the release date of the results: Sept. 15 2020
- associated workshop days: Oct. 04 2020, 9:00-13:00 UTC

Ethics Approval

Training dataset: The ethics approval was granted by the Research Ethics Committee of McGill University Health Center (Protocol Number: MM-JGH-CR15-50). Test dataset: The ethics approval was obtained from the Commission cantonale (VD) d'éthique de la recherche sur l'étre humain (CER-VD) with protocol number: 2018-01513.

---

[11] www.aicrowd.com/challenges/hecktor#results-submission%20format.

Data Usage Agreement
The participants had to fill out and sign an end-user-agreement in order to be granted access to the data. The form can be found under the Resources tab of the HECKTOR website.

Code Availability
The evaluation software was made available on our github page[12]. The participating teams decided whether they wanted to disclose their code (they were encouraged to do so).

Conflict of Interest
No conflict of interest applies. Fundings are specified in the acknowledgments. Only the organizers had access to the test cases ground truth contours.

Author Contributions

Vincent Andrearczyk
Design of the task and of the challenge, writing of the proposal, development of baseline algorithms, development of the AIcrowd website, writing of the overview paper, organization of the challenge event, organization of the submission and reviewing process of the participants' papers.

Valentin Oreiller
Design of the task and of the challenge, writing of the proposal, development of the AIcrowd website, development of the evaluation code, writing of the overview paper, organization of the challenge event, organization of the submission and reviewing process of the papers.

Mario Jreige
Design of the task and of the challenge, quality control/annotations, annotations for inter-annotator agreement, revision of the paper and accepted the last version of the submitted paper.

Martin Vallières
Design of the task and of the challenge, provided the initial data and annotations for the training set [20], revision of the paper and accepted the last version of the submitted paper.

Joel Castelli
Design of the task and of the challenge, annotations for inter-annotator agreement.

Hesham Elhalawani
Design of the task and of the challenge, annotations for inter-annotator agreement.

Sarah Boughdad
Design of the task and of the challenge, annotations for inter-annotator agreement.

---

[12] github.com/voreille/hecktor/tree/master/src/evaluation.

John O. Prior
Design of the task and of the challenge, revision of the paper and accepted the last version of the submitted paper.

Adrien Depeursinge
Design of the task and of the challenge, writing of the proposal, writing of the overview paper, organization of the challenge event.

Appendix 2: Image Acquisition Details

HGJ: All patients had FDG-PET and CT scans done on a hybrid PET/CT scanner (Discovery ST, GE Healthcare) within 37 days before treatment (median: 14 days). For the PET portion of the FDG-PET/CT scan, a median of 584 MBq (range: 368–715) was injected intravenously. Imaging acquisition of the head and neck was performed using multiple bed positions with a median of 300 s (range: 180–420) per bed position. Attenuation corrected images were reconstructed using an ordered subset expectation maximization (OSEM) iterative algorithm and a span (axial mash) of 5. The FDG-PET slice thickness resolution was 3.27 mm for all patients and the median in-plane resolution was $3.52 \times 3.52 \text{ mm}^2$ (range: 3.52–4.69). For the CT portion of the FDG-PET/CT scan, an energy of 140 kVp with an exposure of 12 mAs was used. The CT slice thickness resolution was 3.75 mm and the median in-plane resolution was $0.98 \times 0.98 \text{ mm}^2$ for all patients.

CHUS: All 102 eligible patients had FDG-PET and CT scans done on a hybrid PET/CT scanner (GeminiGXL 16, Philips) within 54 days before treatment (median: 19 days). For the PET portion of the FDG-PET/CT scan, a median of 325 MBq (range: 165–517) was injected intravenously. Imaging acquisition of the head and neck was performed using multiple bed positions with a median of 150 s (range: 120–151) per bed position. Attenuation corrected images were reconstructed using a LOR-RAMLA iterative algorithm. The FDG-PET slice thickness resolution was 4 mm and the median in-plane resolution was $4 \times 4 \text{ mm}^2$ for all patients. For the CT portion of the FDG-PET/CT scan, a median energy of 140 kVp (range: 12–140) with a median exposure of 210 mAs (range: 43–250) was used. The median CT slice thickness resolution was 3 mm (range: 2–5) and the median in-plane resolution was $1.17 \times 1.17 \text{ mm}^2$ (range: 0.68–1.17).

HMR: All patients had FDG-PET and CT scans done on a hybrid PET/CT scanner (Discovery STE, GE Healthcare) within 60 days before treatment (median: 34 days). For the PET portion of the FDG-PET/CT scan, a median of 475 MBq (range: 227–859) was injected intravenously. Imaging acquisition of the head and neck was performed using multiple bed positions with a median of 360 s (range: 120–360) per bed position. Attenuation corrected images were reconstructed using an ordered subset expectation maximization (OSEM) iterative algorithm and a median span (axial mash) of 5 (range: 3–5). The FDG-PET slice thickness resolution was 3.27 mm for all patients and the median in-plane resolution was $3.52 \times 3.52 \text{ mm}^2$ (range: 3.52–5.47). For the CT portion of the FDG-PET/CT scan, a median energy of 140 kVp (range: 120–140) with a median exposure of 11 mAs (range: 5–16) was used. The CT slice thickness resolution was 3.75 mm for all patients and the median in-plane resolution was $0.98 \times 0.98\,\text{mm}^2$ (range: 0.98–1.37).

CHUM: All patients had FDG-PET and CT scans done on a hybrid PET/CT scanner (Discovery STE, GE Healthcare) within 66 days before treatment (median: 12 days). For the PET portion of the FDG-PET/CT scan, a median of 315 MBq (range: 199–3182) was injected intravenously. Imaging acquisition of the head and neck was performed using multiple bed positions with a median of 300 s (range: 120–420) per bed position. Attenuation corrected images were reconstructed using an ordered subset expectation maximization (OSEM) iterative algorithm and a medianspan (axial mash) of 3 (range: 3–5). The median FDG-PET slice thickness resolution was 4 mm (range: 3.27–4) and the median in-plane resolution was $4 \times 4\,\text{mm}^2$ (range: 3.52–5.47). For the CT portion of the FDG-PET/CT scan, a median energy of 120 kVp (range: 120–140) with a median exposure of 350 mAs (range: 5–350) was used. The median CT slice thickness resolution was 1.5 mm (range: 1.5–3.75) and the median in-plane resolution was $0.98 \times 0.98\,\text{mm}^2$ (range: 0.98–1.37). All patients received their FDG-PET/CT scan dedicated to the head and neck area right before their planning CT scan, in the same position with the immobilization device.

CHUV (test): All patients underwent FDG PET/CT for staging before treatment. Blood glucose levels were checked before the injection of (18F)-FDG. After a 60-min uptake period of rest, patients were imaged with the Discovery D690 TOF PET/CT (General Electric Healthcare, Milwaukee, WI, USA). First, a CT (120 kV, 80 mA, 0.8-s rotation time, slice thickness 3.75 mm) was performed from the base of the skull to the mid-thigh. PET scanning was performed immediately after acquisition of the CT. Images were acquired from the base of the skull to the mid-thigh (2 min/bed position). PET images were reconstructed after time-of-flight and point-spread-function recovery corrections by using an ordered-subset expectation maximization iterative reconstruction (OSEM) (two iterations, 28 subsets) and an iterative fully 3D (Discovery ST). CT data were used for attenuation calculation.

References

1. Andrearczyk, V., Oreiller, V., Depeursinge, A.: Oropharynx detection in PET-CT for tumor segmentation. In: Irish Machine Vision and Image Processing (2020)
2. Andrearczyk, V., et al.: Automatic segmentation of head and neck tumors and nodal metastases in PET-CT scans. In: International Conference on Medical Imaging with Deep Learning (MIDL) (2020)
3. Bogowicz, M., Tanadini-Lang, S., Guckenberger, M., Riesterer, O.: Combined CT radiomics of primary tumor and metastatic lymph nodes improves prediction of loco-regional control in head and neck cancer. Sci. Rep. **9**(1), 1–7 (2019)
4. Chen, H., Chen, H., Wang, L.: Iteratively refine the segmentation of head and neck tumor in FDG-PET and CT images. In: Andrearczyk, V., et al. (eds.) HECKTOR 2020. LNCS, vol. 12603, pp. 53–58. Springer, Cham (2021)

5. Foster, B., Bagci, U., Mansoor, A., Xu, Z., Mollura, D.J.: A review on segmentation of positron emission tomography images. Comput. Biol. Med. **50**, 76–96 (2014)
6. Ghimire, K., Chen, Q., Feng, X.: Patch-based 3D UNet for head and neck tumor segmentation with an ensemble of conventional and dilated convolutions. In: Andrearczyk, V., et al. (eds.) HECKTOR 2020. LNCS, vol. 12603, pp. 78–84. Springer, Cham (2021)
7. Gudi, S., et al.: Interobserver variability in the delineation of gross tumour volume and specified organs-at-risk during IMRT for head and neck cancers and the impact of FDG-PET/CT on such variability at the primary site. J. Med. Imaging Radiat. Sci. **48**(2), 184–192 (2017)
8. Hatt, M., et al.: The first MICCAI challenge on PET tumor segmentation. Med. Image Anal. **44**, 177–195 (2018)
9. Hatt, M., et al.: Classification and evaluation strategies of auto-segmentation approaches for PET: report of AAPM task group no. 211. Med. Phys. **44**(6), e1–e42 (2017)
10. Iantsen, A., Visvikis, D., Hatt, M.: Squeeze-and-excitation normalization for automated delineation of head and neck primary tumors in combined PET and CT images. In: Andrearczyk, V., et al. (eds.) HECKTOR 2020. LNCS, vol. 12603, pp. 37–43. Springer, Cham (2021)
11. Kumar, A., Fulham, M., Feng, D., Kim, J.: Co-learning feature fusion maps from PET-CT images of lung cancer. IEEE Trans. Med. Imaging **39**(1), 204–217 (2019)
12. Li, L., Zhao, X., Lu, W., Tan, S.: Deep learning for variational multimodality tumor segmentation in PET/CT. Neurocomputing **392**, 277–295 (2019)
13. Ma, J., Yang, X.: Combining CNN and hybrid active contours for head and neck tumor segmentation in CT and PET Images. In: Andrearczyk, V., et al. (eds.) HECKTOR 2020. LNCS, vol. 12603, pp. 59–64. Springer, Cham (2021)
14. Maier-Hein, L., et al.: Why rankings of biomedical image analysis competitions should be interpreted with care. Nat. Commun. **9**(1), 1–13 (2018)
15. Maier-Hein, L., et al.: BIAS: transparent reporting of biomedical image analysis challenges. Med. Image Anal. **66**, 101796 (2020)
16. Moe, Y.M., et al.: Deep learning for automatic tumour segmentation in PET/CT images of patients with head and neck cancers. In: Medical Imaging with Deep Learning (2019)
17. Naser, M.A., van Dijk, L.V., He, R., Wahid, K.A., Fuller, C.D.: Tumor segmentation in patients with head and neck cancers using deep learning based-on multimodality PET/CT images. In: Andrearczyk, V., et al. (eds.) HECKTOR 2020. LNCS, vol. 12603, pp. 85–98. Springer, Cham (2021)
18. Rao, C., et al.: Oropharyngeal Tumour Segmentation using Ensemble 3D PET-CT Fusion Networks for the HECKTOR Challenge. In: Andrearczyk, V., et al. (eds.) HECKTOR 2020. LNCS, vol. 12603, pp. 65–77. Springer, Cham (2021)
19. Ronneberger, O., Fischer, P., Brox, T.: U-Net: convolutional networks for biomedical image segmentation. In: Navab, N., Hornegger, J., Wells, W.M., Frangi, A.F. (eds.) MICCAI 2015. LNCS, vol. 9351, pp. 234–241. Springer, Cham (2015). https://doi.org/10.1007/978-3-319-24574-4_28
20. Vallieres, M., et al.: Radiomics strategies for risk assessment of tumour failure in head-and-neck cancer. Sci. Rep. **7**(1), 1–14 (2017)
21. Xie, J., Peng, Y.: The head and neck tumor segmentation using nnU-Net with spatial and channel 'squeeze & excitation' blocks. In: Andrearczyk, V., et al. (eds.) HECKTOR 2020. LNCS, vol. 12603, pp. 28–36. Springer, Cham (2021)

22. Xu, L., et al.: Automated whole-body bone lesion detection for multiple myeloma on 68Ga-pentixafor PET/CT imaging using deep learning methods. Contrast Media Mol. Imaging **2018**, 11 (2018). https://doi.org/10.1155/2018/2391925
23. Yousefirizi, F., Rahmim, A.: GAN-based bi-modal segmentation using mumford-shah loss: Application to head and neck tumors in PET-CT images. In: Andrearczyk, V., et al. (eds.) HECKTOR 2020. LNCS, vol. 12603, pp. 99–108. Springer, Cham (2021)
24. Yuan, Y.: Automatic head and neck tumor segmentation in PET/CT with scale attention network. In: Andrearczyk, V., et al. (eds.) HECKTOR 2020. LNCS, vol. 12603, pp. 44–52. Springer, Cham (2021)
25. Zhao, X., Li, L., Lu, W., Tan, S.: Tumor co-segmentation in PET/CT using multi-modality fully convolutional neural network. Phys. Med. Biol. **64**(1), 015011 (2018)
26. Zhong, Z., et al.: 3D fully convolutional networks for co-segmentation of tumors on PET-CT images. In: 2018 IEEE 15th International Symposium on Biomedical Imaging (ISBI 2018), pp. 228–231. IEEE (2018)
27. Zhu, S., Dai, Z., Ning, W.: Two-stage approach for segmenting gross tumor volume in head and neck cancer with CT and PET imaging. In: Andrearczyk, V., et al. (eds.) HECKTOR 2020. LNCS, vol. 12603, pp. 22–27. Springer, Cham (2021)

Two-Stage Approach for Segmenting Gross Tumor Volume in Head and Neck Cancer with CT and PET Imaging

Simeng Zhu[✉], Zhenzhen Dai, and Ning Wen

Department of Radiation Oncology, Henry Ford Cancer Institute,
Detroit, MI, USA
szhu1@hfhs.org

Abstract. Radiation treatment planning for head and neck cancers involves careful delineation of the tumor target volume on CT images, often with assistance from PET scans as well. In this study, we described a method to automatically segment the gross tumor volume of the primary tumor with a two-stage approach using deep convolutional neural networks as part of the HECKTOR challenge. We trained a classification network to select the axial slices which may contain the tumor, and these slices were then inputted into a segmentation network to generate a binary segmentation map. On the test set consisting of 53 patients, we achieved a mean Dice similarity coefficient of 0.644, mean precision of 0.694, and mean recall of 0.667.

Keywords: Tumor segmentation · Head and neck cancer · Convolutional neural network

1 Introduction

Radiation therapy is a common and effective treatment modality for head and neck cancers, whether as the primary or adjuvant treatment. Modern radiation therapy allows the delivery of high dose of radiation dose to the tumor and other regions of high risk of containing tumor cells while minimizing radiation dose to the surrounding normal tissues. This is achieved by careful delineation of the tumor target volume and organs-at-risk on computed tomography (CT) images by a radiation oncologist, often with additional information provided by positron emission tomography (PET) images. The delineation of the gross tumor volume (GTVt) of the primary tumor, the visible tumor on imaging, is an important task in this process.

In the past few years, there has been a strong interest in the application of deep convolutional neural networks (CNN) in various computer vision tasks in medical imaging, including segmentation. The HECKTOR (HEad and neCK TumOR segmentation) challenge is such an attempt which aims to develop an automatic segmentation algorithm for the GTVt in head and neck cancers [1, 2]. This challenge is important in two aspects: (1) it realistically simulates the task faced by radiation

© Springer Nature Switzerland AG 2021
V. Andrearczyk et al. (Eds.): HECKTOR 2020, LNCS 12603, pp. 22–27, 2021.
https://doi.org/10.1007/978-3-030-67194-5_2

oncologists by including both CT and FDG-PET imaging as input, and (2) due to the complex anatomical structures, head and neck cancer is known to be one of the most difficult and time-consuming disease sites for contouring within the radiation oncology community, and the development of such an algorithm may potentially increase the radiation oncologist's efficiency in the clinic.

In this paper, we describe our team's method in tackling this challenge. We developed a two-stage deep neural network by assembling a modified ResNet and modified U-Net for tumor segmentation. The modified ResNet was used for classification to select the axial slices which may contain GTVt and the modified U-Net for segmenting the GTVt in the axial slices which were selected by the previous network.

2 Materials

2.1 Dataset Description

The training dataset consists of 201 patients from four hospitals. For each patient, the input consists of co-registered 3D CT image (in Hounsfield unit, HU) and FDG-PET image (in standard uptake value, SUV), and the label is a CT resolution-based 3D binary distribution (of 0's and 1's) for the GTVt. Voxel sizes vary among patients because they were acquired from different institutions. The voxels sizes were about 1.0 mm in the x and y directions but varied from 1.5 to 3.0 mm along the z direction. A 144 mm $\times$ 144 mm $\times$ 144 mm bounding box was provided, indicating the region containing the oropharyngeal tumor.

It is worth noting that, since the SUV on PET images indicates the amount of metabolic activity, the areas that are expected to have elevated SUVs include the primary tumor, tumor-involved lymph nodes, and other normal tissues/organs with normally hypermetabolic activity, but the goal of this challenge is to segment the primary tumor (GTVt) alone while excluding tumor-involved lymph nodes (GTVn).

The test dataset consists of 53 patients from a single hospital. The input data provided for each patient is similar to the training dataset, and no segmentation mask is provided.

2.2 Data Preprocessing

The original 3D CT and PET images were registered and resampled to 1 mm $\times$ 1 mm $\times$ 1 mm pixel spacing with bilinear interpolation. We then cropped an isotropic volume of size 144 mm $\times$ 144 mm $\times$ 144 mm, which was randomly shifted by up to 10 mm in x and y directions from the provided bounding box for data augmentation to avoid model overfitting during training.

The CT volumes were clipped to the [−150, 150] HU range, a common setting for the head and neck region where the tissues of interest were mainly soft tissue, fluid, and fat [1]. In addition, each CT volume was further standardized by division of the standard deviation of the HUs of all voxels within the cropped volume to minimize inter-scanner effects. Inspired by preprocessing method used in previous studies [3], Each PET volume was clipped to the $[0.2 * SUV_{max}, 10.0]$ SUV range, followed by a division of 10.0 to keep the orders of magnitude similar between the values for CT and PET volumes.

We then sliced the CT and PET volumes into 2D axial slices and concatenated the corresponding CT and PET slices as 2 channels such that, for each patient, there were 144 2D slices of size (144, 144, 2).

3 Methods

3.1 Overall Network Scheme

We used a two-stage approach by separately training a classification CNN and a segmentation U-Net. The classification network selects a series of 2D slices containing GTVt, and the segmentation network subsequently predicts a segmentation map based on these slices. The overall scheme is demonstrated in Fig. 1.

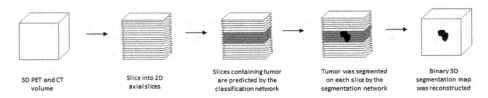

| 3D PET and CT volume | Slice into 2D axial slices | Slices containing tumor are predicted by the classification network | Tumor was segmented on each slice by the segmentation network | Binary 3D segmentation map was reconstructed |

Fig. 1. An overview of our method.

3.2 Classification Network

The goal of the classification network is to select axial slices that may contain the GTVt such that the segmentation network can focus on segmentation. The architecture of this ResNet-inspired network is shown in Fig. 2: an input slice of size (144, 144, 2) first undergoes a 7 × 7 Conv-BatchNorm-ReLU-MaxPool block, followed by a series of 3 × 3 Conv-BatchNorm-ReLU blocks with residual units [4]. Eventually, a 3 × 3 × 1 feature map undergoes average pooling and sigmoid operation to produce a number in the range [0.0, 1.0], corresponding to the likelihood of the slice containing GTVt. The classification network was trained with all axial slices from the training set. Binary cross entropy for the overall accuracy was as the loss function during training. Further information on training is detailed in Sect. 4.1.

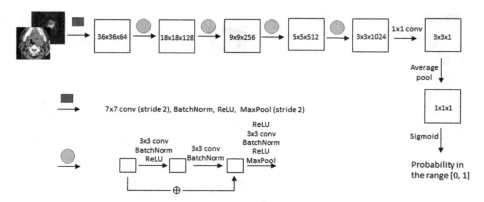

Fig. 2. Details of the ResNet-inspired classification network (the dimension of feature maps is reported as width × height × number of channels).

3.3   Segmentation Network

The goal of the segmentation network is to output a segmentation map given a 2D slice of high likelihood of containing GTVt. As shown in Fig. 3, the network architecture resembles U-net which employs encoder-decoder with skip connections to facilitate improved gradient flow during backpropagation [5]. In addition, other features including attention gating, dilated convolution, and residual units were also incorporated, as they have been shown to improve segmentation performance of U-Net in prior studies. Oktay et al. demonstrated that the addition of attention gating in U-Net significantly improved the performance of pancreas segmentation on CT scans [6], and this feature might be helpful in our study for the model to efficiently identify the tumor based on CT features and PET avidity. Dilated convolutions have been shown to be useful in expanding the receptive field for the network [7], which is important for the task in this study as the network needs to learn to ignore PET avidity in locations where head and neck cancers cannot occur (e.g. the brain). In addition, residual units are used to allow additional gradient flow during backpropagation in addition to the skip connections [4].

The segmentation network was trained with GTVt-containing slices only. The soft Dice loss was used as the loss function during training. Further information on training is detailed in Sect. 4.1.

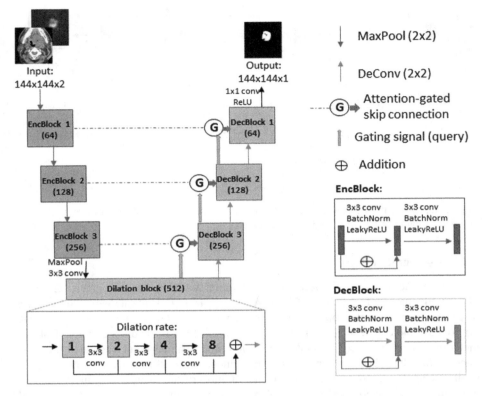

Fig. 3. Details of the attention-gated, dilated, residual U-Net for segmentation. Numbers in parentheses indicate the number of feature maps.

4   Experiment

4.1   Training Details

The total of 201 cases were randomly divided into 181 for training and 20 for validation; therefore, there were 26,004 axial 2D slices of CT and PET images available for training, of which 6,379 slices contained GTVt.

The classification network was trained on both slices that contain GTVt ("positive slices") and those that do not ("negative slices"). However, due to data imbalance between the two classes, we under-sampled the negative slices during training; therefore, each epoch contains all 6,379 "positive slices" and 6,379 randomly sampled "negative slices". Batch size was 10. Adam was used as the optimizer, with a learning rate of 0.005, β_1 of 0.9, β_2 of 0.999, and epsilon of 10^{-7}. Training was stopped when the overall classification accuracy of the validation set decreased for 2 consecutive epochs to reduce overfitting. During inference, a threshold of 0.5 was used for classification.

The segmentation network was trained using "positive slices" only. Adam was used as the optimizer, with a learning rate of 0.005, β_1 of 0.9, β_2 of 0.999, and epsilon of 10^{-7}. Batch size was 1. The network was trained for 20 epochs, and the model from the epoch with the highest Dice similarity coefficient (DSC) on the validation set was used for inference to avoid overfitting. During inference, a threshold of 0.5 across all pixels was used for segmentation.

The program was implemented in Python 3.8 with Tensorflow 2.2 as the framework. The network training was performed on both Google Colab and NVIDIA Tesla V100 GPU.

4.2 Results

The best performing classification and segmentation models on the validation dataset were independently selected and combined to build the final model. With the entire 3D volume as the model input, the final model achieved a mean DSC of 0.644, mean precision of 0.694, and mean recall of 0.667 in the test set consisting of 53 patients.

5 Conclusion

We described a two-stage convolutional neural network-based method for automatic segmentation of gross tumor volume of the primary tumor in head and neck cancer. The input images first go through a classification network to predict the presence of the tumor and, if so, goes through a second the segmentation network to produce a binary segmentation map. We have achieved excellent Dice similarity coefficient with this approach.

References

1. Andrearczyk, V., et al.: Automatic segmentation of head and neck tumors and nodal metastases in PET-CT scans. In: Proceedings of Machine Learning Research, pp. 1–11 (2020)
2. Andrearczyk, V., et al.: Overview of the HECKTOR challenge at MICCAI 2020: automatic head and neck tumor segmentation in PET/CT. In: Andrearczyk, V., et al. (eds.) HECKTOR 2020. LNCS, vol. 12603, pp. 1–21. Springer, Cham (2021)
3. Gsaxner, C., Roth, P.M., Wallner, J., Egger, J.: Exploit fully automatic low-level segmented PET data for training high-level deep learning algorithms for the corresponding CT data. PLoS ONE 14(3), e0212550 (2019)
4. He, K., Zhang, X., Ren, S., Sun, J.: Deep residual learning for image recognition. In: 2016 IEEE Conference on Computer Vision and Pattern Recognition, pp. 770–778 (2016)
5. Ronneberger, O., Fischer, P., Brox, T.: U-Net: convolutional networks for biomedical image segmentation. In: Navab, N., Hornegger, J., Wells, W.M., Frangi, A.F. (eds.) MICCAI 2015. LNCS, vol. 9351, pp. 234–241. Springer, Cham (2015). https://doi.org/10.1007/978-3-319-24574-4_28
6. Schlemper, J., et al.: Attention-gated networks: learning to leverage salient regions in medical images. Med. Image Anal. 53, 197–207 (2019)
7. Devalla, S.K., et al.: DRUNET: a dilated-residual U-Net deep learning network to segment optic nerve head tissues in optical coherence tomography images. Biomed. Opt. Express 9(7), 3244–3265 (2018)

The Head and Neck Tumor Segmentation Using nnU-Net with Spatial and Channel 'Squeeze & Excitation' Blocks

Juanying Xie[(✉)] and Ying Peng

School of Computer Science, Shaanxi Normal University, Xi'an 710119,
People's Republic of China
xiejuany@snnu.edu.cn

Abstract. The head and neck (H&N) cancer is the eighth most common cause of cancer death. Radiation therapy is one of the most effective therapies, but it heavily relies on the contouring of tumor volumes on medical images. In this paper, the 3D nnU-Net is first applied to segment H&N tumors in FluoroDeoxyGlucose Positron Emission Tomography (FDG-PET) and Computed Tomography (CT) images. Furthermore we improve upon the 3D nnU-Net by integrating it the spatial and channel 'squeeze & excitation' (scSE) blocks, so as to boost those meaningful features while suppressing weak ones. We name the advanced 3D nnU-Net as 3D scSE nnU-Net. Its performance is tested on the HECKTOR 2020 training data by dividing it into training and validation subsets, such as 160 images are in training subset and 41 images are contained in validation subset. The experimental results on the validation images show that the proposed 3D scSE nnU-Net is superior to the original 3D nnU-Net by 1.4% in terms of DSC (Dice Similarity Coefficient) metric on this segmentation task. Our 3D scSE nnU-Net has got the DSC of 0.735 on HECKTOR test data. It has got the third place in this HECKTOR challenge.

Keywords: Head and neck cancer · nnU-Net · Squeeze & excitation · Semantic segmentation

1 Introduction

Head and neck (H&N) cancer is the fifth most common cancer diagnosed worldwide and the eighth most common cause of cancer death [1]. Radiation therapy is one of the most effective therapies. It relies heavily on the contouring of tumor volumes on medical images. This task is usually done manually, so it is usually a time-consuming and labour-intensive task. It is prone to bring the inter- and intra-observer differences [2]. Therefore, it makes sense to design computer-aided systems that can automatically segment the target area of H&N cancer, so as to reduce the workload of radiologists while reducing the variety of boundaries drawn by different observers.

Segmentation of H&N tumors is a more difficult task than the segmentations of the images of any other parts of the body. The main reason is that the density values of H&N tumors are very similar to those of adjacent tissues, such that it cannot be distinguished by naked eyes, especially in CT images [3].

© Springer Nature Switzerland AG 2021
V. Andrearczyk et al. (Eds.): HECKTOR 2020, LNCS 12603, pp. 28–36, 2021.
https://doi.org/10.1007/978-3-030-67194-5_3

Head and Neck Tumor Segmentation Challenge (HECKTOR) aims to advance and test the methodologies to realize the automatic tumor segmentation in PET/CT images by providing a 3D PET/CT dataset with ground truth tumor segmentation labels annotated by physicians [4]. PET-CT has played an important role in the diagnosis and treatment of H&N tumors. It can provide both anatomical and metabolic information about the tumor. HECKTOR 2020 training data comprise 201 cases, each with two 3D modalities (PET, CT). The data are collected from four institutions, using various hybrid PET/CT scanners. The test data (53 cases) are from different institutions and are used to calculate the final challenge ranking.

It is well known that deep learning methods have been demonstrated as a technology with high potential and have achieved the state-of-the-art results in many medical segmentation tasks in the past decade. The nnU-Net [5] is the best baseline among those deep learning methods for medical segmentation tasks. Therefore, we adopt its idea to solve the HECKTOR 2020 task. We adopt the encoder-decoder structure of nnU-Net while integrating the scSE blocks [6] to the nnU-Net network to aggregate its global information. We describe our pipeline in detail while introducing data used in this paper briefly in the Sect. 2. Section 3 shows the results. Discussions to the results are in Sect. 4. Conclusions come in Sect. 5.

2 Data and Methods

This section first introduce the data of HECKTOR 2020 briefly. Then it describes our methods in detail. We describe the data preprocessing method we used in this paper. We also summarize the 3D nnU-Net published in [5]. Then we introduce our proposed 3D scSE nnU-Net. Followed that we describe the training scheme we used in our experiments. The main idea is that we adopt the 3D nnU-Net in [5] as the baseline, then advance it to the 3D scSE nnU-Net in this paper, so as to advance the performance of the 3D nnU-Net on the HECKTOR 2020 task.

2.1 Data

The data used in the experiments in this paper are from HECKTOR 2020. The training data comprise 201 cases from four centers of CHGJ, CHMR, CHUM and CHUS. The test data comprise 53 cases from another center of CHUV. We can get the training data when we were taking part in the challenge. Each training case comprises CT, PET and GTVt (primary Gross Tumor Volume, that is, the label of each case) in NIfTI format, as well as the bounding box location and patient information in CSV format file. The test data are without GTVt information.

We divide the HECKTOR 2020 training data into training and validation subsets under the condition that the original distribution of training data is preserved. The training subset comprises 160 training images from 201 training cases. The validation subset includes 41 validation images from 201 training cases. The training images are used to train the model, and the validation images are to estimate the performance of the model.

2.2 Data Preprocessing

To reduce the amount of calculations, we first crop all images in the training data to the region of nonzero values. That is, the same edge region of the images of one patient will be cropped if the values of the edge region are zeroes in both CT and PET images. Secondly, due to the fact that different scanners or different acquisition protocols may result in data with heterogeneous voxel spacings, we resample all cases to a median voxel spacing of all CT images, such as $0.9765625 \times 0.9765625 \times 3.0$ mm, so as to result in a median image shape of $147 \times 147 \times 48$ voxels for the HECKTOR training data. Finally, in order to accelerate the neural network to converge when training it, image intensities are normalized. For CT images, all foreground voxels of the CT images in the HECKTOR training data are collected and an automated level-window-like cropping of intensity values is performed based on the 0.5th and 99.5th percentile of these values [5]. For PET, we normalize the intensity values by using z-score normalization.

2.3 Baseline: 3D nnU-Net

The architecture of the 3D nnU-Net is shown in Fig. 1. The input image size of the 3D nnU-Net is $2 \times 48 \times 160 \times 160$. If the size of the cropped image is smaller than this input image size, the border is filled with 0. The 3D nnU-Net uses two conv-instnorm-leaky ReLU blocks between downsamplings and upsamplings in both encoder and decoder. Downsampling is done via strided convolution and upsampling is finished with transposed convolution. This architecture initially uses 30 feature maps, which is doubled for each downsampling operation in the encoder (up to 320 feature maps) and halved for each transposed convolution in the decoder [5]. The end of the decoder has the same spatial size as the input, followed by a $1 \times 1 \times 1$ convolution into 2 channels and a softmax function.

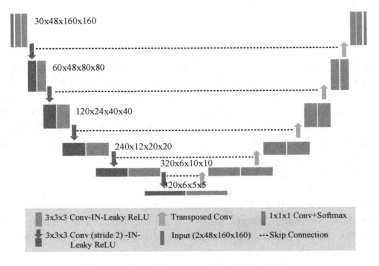

Fig. 1. The architecture of the 3D nnU-Net.

2.4   The 3D ScSE nnU-Net

The 3D U-Net [7] has served as the backbone network for medical image segmentation due to its good performance. However, the upsampling process involves the recovery of spatial information, which is hard without taking the global information into consideration [8]. The scSE blocks are able to recalibrate the directions of the learned feature maps adaptively to boost meaningful features and suppress weak ones. Therefore, we integrate the scSE blocks into the 3D nnU-Net by following its encoder and decoder blocks. The new proposed architecture of the 3D scSE nnU-Net is shown in Fig. 2.

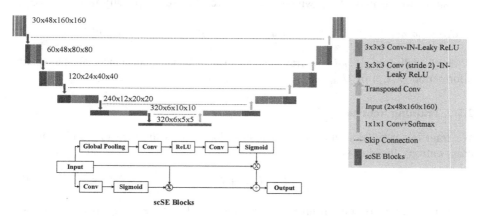

Fig. 2. The architecture of the proposed 3D scSE nnU-Net.

This scSE blocks contain two blocks, and they are cSE and sSE. For the cSE block, we represent an intermediate input feature map $U = [u_1, u_2, \cdots, u_C]$ as a combination of channels $u_m \in R^{Z \times H \times W}$, where Z, H and W are the depth, height and width of U, respectively. Spatial squeeze is done by a global average pooling layer, so as to produce the vector $z \in R^{C \times 1 \times 1 \times 1}$. Its mth element is calculated in (1).

$$z_m = \frac{1}{Z \times H \times W} \sum_{k}^{Z} \sum_{i}^{H} \sum_{j}^{W} u_m(k, i, j) \tag{1}$$

Where $u_m(k, i, j)$ corresponds to the value of the mth channel of U at the spatial location (k, i, j), with $k \in \{1, 2, \cdots, Z\}$, $i \in \{1, 2, \cdots, H\}$ and $j \in \{1, 2, \cdots, W\}$.

This vector z embeds global spatial information. It is transformed into $\hat{z} = W_2(\delta(W_1 z))$, where $W_1 \in R^{\frac{C}{r} \times C \times 1 \times 1 \times 1}$ and $W_2 \in R^{C \times \frac{C}{r} \times 1 \times 1 \times 1}$ are the weights of two convolution layers, respectively, and $\delta(\cdot)$ is the ReLU operator. The parameter r in W_1 represents the bottleneck of the channel excitation, which encodes the channel-wise dependencies. In our experiments, we set $r = 2$. Finally, $\hat{z}$ is passed through a sigmoid layer $\sigma(\hat{z})$. The resultant vector indicates the importance of the mth channel, which is either scaled up or down, and is used to recalibrate U to $\hat{U}_{cSE}$ in (2).

$$\hat{U}_{cSE} = [\sigma(\hat{z}_1)u_1, \sigma(\hat{z}_2)u_2, \cdots, \sigma(\hat{z}_C)u_C] \tag{2}$$

For the sSE block, we consider an alternative slicing of the input feature map $U = [u(1,1,1), u(1,1,2), \cdots, u(k,i,j), \cdots, u(Z,H,W)]$, where $u(k,i,j) \in R^{C \times 1 \times 1 \times 1}$. The spatial squeeze operation is done by convolution $q = W_{sq}U$ with $W_{sq} \in R^{1 \times C \times 1 \times 1 \times 1}$, generating a projection tensor $q \in R^{Z \times H \times W}$. Each $q(k,i,j)$ of the projection represents the linearly combined representation for all channels C for the spatial location (k,i,j). Then, q is passed through a sigmoid layer $\sigma(\cdot)$ to rescale activations to fall into [0, 1], which is used to recalibrate U spatially to $\hat{U}_{sSE}$ in (3).

$$\hat{U}_{sSE} = [\sigma(q(1,1,1))u(1,1,1), \cdots, \sigma(q(k,i,j))u(k,i,j), \cdots, \sigma(q(Z,H,W))u(Z,H,W)] \tag{3}$$

Each value $\sigma(q(k,i,j))$ corresponds to the relative importance of the spatial information (k,i,j) of U.

2.5   Training Scheme

We train our networks with the combination loss that is the dice loss and weighted cross-entropy loss in (4).

$$L = L_{dice} + L_{WCE} \tag{4}$$

This dual loss benefits from both the smooth and bounded gradients of the cross-entropy loss and the explicit optimization of the Dice score used for evaluation and its robustness to class imbalance. The dice loss used here is a multi-class variation loss proposed in [9]. It is calculated as follows in (5).

$$L_{dice} = -\frac{2}{|T|} \sum_{t \in T} \frac{\sum_{i \in I} u_i^t v_i^t}{\sum_{i \in I} u_i^t + \sum_{i \in I} v_i^t} \tag{5}$$

Where, u is the softmax output of the network and v is the one hot encoding of the ground truth. Both u and v have the shape of $I \times T$, with $i \in I$, and $t \in T$. Here I is the number of voxels in the training patch/batch, and T is the number of classes.

The weighted cross-entropy loss L_{WCE} is calculated as follows in (6). In our experiments we set w = 3.0.

$$L_{WCE} = -\sum wv_i \log u_i + (1 - v_i) \log(1 - u_i) \tag{6}$$

We use stochastic gradient descent optimizer to train our networks. The initial learning rate is $\alpha_0 = 1e - 3$. The learning rate is updated in (7).

$$\alpha = \alpha_0 * (1 - \frac{e}{N_e})^{0.9} \tag{7}$$

Where e is an epoch counter, N_e is a total number of epochs. We set $N_e = 1000$ in our experiments. We use a batch size of 2. The model converges after 250 epochs. We also regularize the convolutional kernel parameters using the $L2$ norm regularization with a weight of $5e - 5$.

In addition, to prevent overfitting when training a large neural network using the limited training data, we adopt the following augmentation techniques, such as random rotations, random scaling, random elastic deformations, gamma correction augmentation and mirroring, to enlarge the training subset data during training process.

We implement our network in Pytorch and trained it on a single GeForce RTX 2080Ti GPU with 11 GB memory. The training of a single network is about 10 h.

3 Results

This section shows all of the experimental results we obtained. Because the HECKTOR 2020 competition uses the Dice Similarity Coefficient (DSC) as the evaluation metrics, so we take it as the metric to value performance of the model we trained. The quantitative results are summarized in Table 1 in terms of DSC. Table 2 compares the average inference time for validation image in the validation subset using our 3D scSE nnU-Net and the original 3D nnU-Net. Table 3 displays the top 5 results on test data of competition teams for HECKTOR 2020 challenge.

To further demonstrate the effectiveness of our proposed 3D scSE nnU-Net, we compare its training loss function curve to that of the 3D nnU-Net in Fig. 3, and also show several segmentation results on validation data of our 3D scSE nnU-Net and of the original 3D nnU-Net in Fig. 4.

Table 1. The mean segmentation results in terms of DSC on validation subset.

Model	DSC metric
3D nnU-Net	0.812
3D scSE nnU-Net	0.826

Table 2. The average inference time in seconds of images from validation subset.

Model	Inference time/s
3D nnU-Net	1.83
3D scSE nnU-Net	2.64

Table 3. The results on test data of Top 5 ranks in HECKTOR 2020 challenge.

Team (Rank)	DSC metric
Andrei Iantsen (1)	0.759
Jun Ma (2)	0.752
Yading Yuan (4)	0.732
Huai Chen et al. (5)	0.724
Ying Peng et al. (**ours**) (3)	0.735

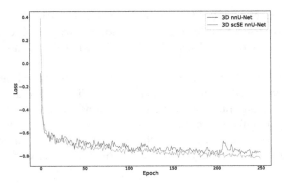

Fig. 3. The training loss function curves of 3D scSE nnU-Net and 3D nnU-Net.

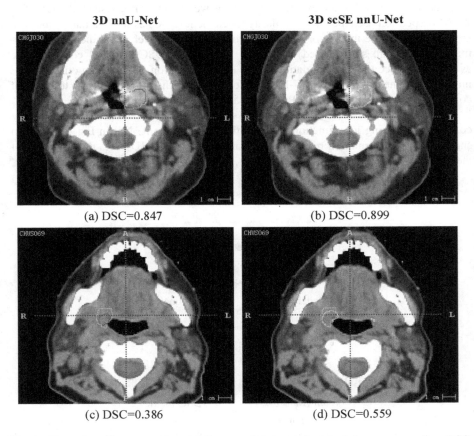

Fig. 4. Examples of segmentation results on validation subset and the corresponding DSC values. Green line represents the ground truth. Purple and yellow lines represent the segmentation results of 3D nnU-Net and 3D scSE nnU-Net, respectively. The top row is an example with good predictive results, and the bottom row is an example with bad predictive results. (Color figure online)

4   Discussions

The results in Table 1 show that the performance of our 3D scSE nnU-Net is superior to that of 3D nnU-Net in terms of DSC metric. The results in Table 2 show that the average inference time of the 3D scSE nnU-Net is a little longer than that of original 3D nnU-Net. This means that we bear a loss in inference time, so as to get the better target detection. It is usually a fact that we always do a tradeoff between efficiency and performance.

The results in Fig. 3 show that our 3D scSE nnU-Net converges faster than the original 3D nnU-Net when training the model. The results in Fig. 4 show that the targets detected by our proposed 3D scSE nnU-Net are much more similar to the actual ones than that by 3D nnU-Net. The DSC values of our 3D scSE nnU-Net are much higher than the corresponding ones of the original 3D nnU-Net.

The Top 5 segmentation results on test data in Table 3 show that our 3D scSE nnU-Net model ranks in the third place in this HECKTOR 2020 challenge. Although the top 1 team can get the DSC value of 0.759, there is still a large gap between it and 0.8. Therefore, it need further study to develop much more advanced method to realize the head and neck tumor segmentation task effectively. This type of study will benefit human being by reducing the death rate of head and neck cancers.

5   Conclusions

In this work, we proposed a semantic segmentation network named as 3D scSE nnU-Net for H&N tumor segmentation task from multimodal 3D images. This 3D scSE nnU-Net advances the original 3D nnU-Net. It realizes the automatic segmentation to the H&N tumors by integrating scSE blocks to the original 3D nnU-Net.

Although the proposed 3D scSE nnU-Net outperforms the original for H&N tumor segmentation task from multimodal 3D images, it needs a little more inference time compared to the original 3D nnU-Net. We will try to advance the performance of our 3D scSE nnU-Net in segmenting the head and neck tumors while preserving its reference time. Furthermore, we will investigate the impact of different data partitions on the model performance, such as leave-one-center-out as test data. We hope we can further carry out a thorough study in the head and neck tumor segmentation task after the HECKTOR 2020 challenge.

Acknowledgements. This work is supported in part by the National Natural Science Foundation of China under grant No. 61673251, 62076159 and 12031010, and is also by the National Key Research and Development Program of China under grant No. 2016YFC0901900, and by the Fundamental Research Funds for the Central Universities under grant No. GK201701006 and 2018TS078, the Scientific and Technological Achievements Transformation and Cultivation Funds under grant No. GK201806013, and the Innovation Funds of Graduate Programs at Shaanxi Normal University under grant No. 2015CXS028 and 2016CSY009.

We also acknowledge the HECKTOR2020 challenge organization committee for their providing the competition platform and inviting us submitting this paper for our success wining the third place in this competition.

References

1. O'rorke, M.A., Ellison, M.V., Murray, L.J., Moran, M., James, J., Anderson, L.A.: Human papillomavirus related head and neck cancer survival: a systematic review and meta-analysis. Oral Oncol. 48(12), 1191–1201 (2012). https://doi.org/10.1016/j.oraloncology.2012.06.019
2. Gudi, S., et al.: Interobserver variability in the delineation of gross tumor volume and specified organs-at-risk during IMRT for head and neck cancers and the impact of FDG-PET/CT on such variability at the primary site. J. Med. Imaging Radiat. Sci. 48(2), 184–192 (2017). https://doi.org/10.1016/j.jmir.2016.11.003
3. Andrearczyk, V., et al.: Automatic segmentation of head and neck tumors and nodal metastases in PET-CT scans. In: Medical Imaging with Deep Learning (MIDL) (2020)
4. Andrearczyk, V., et al.: Overview of the HECKTOR challenge at MICCAI 2020: automatic head and neck tumor segmentation in PET/CT. In: Andrearczyk, V., et al. (eds.) HECKTOR 2020. LNCS, vol. 12603, pp. 1–21. Springer, Cham (2021)
5. Isensee, F., Petersen, J., Kohl, S.A.A., Jäger, P.F., Maier-Hein, K.H.: nnU-Net: breaking the spell on successful medical image segmentation. arXiv preprint arXiv:1904.08128 (2019)
6. Roy, A.G., Navab, N., Wachinger, C.: Recalibrating fully convolutional networks with spatial and channel "squeeze and excitation" blocks. IEEE Trans. Med. Imaging 38(2), 540–549 (2018). https://doi.org/10.1109/TMI.2018.2867261
7. Çiçek, Ö., Abdulkadir, A., Lienkamp, S.S., Brox, T., Ronneberger, O.: 3D U-Net: learning dense volumetric segmentation from sparse annotation. In: Ourselin, S., Joskowicz, L., Sabuncu, M.R., Unal, G., Wells, W. (eds.) MICCAI 2016. LNCS, vol. 9901, pp. 424–432. Springer, Cham (2016). https://doi.org/10.1007/978-3-319-46723-8_49
8. Wang, Z., Zou, N., Shen, D., Ji, S.: Non-local U-Nets for biomedical image segmentation. In: Proceedings of the 34th AAAI Conference on Artificial Intelligence, California, pp. 6315–6322. AAAI Press (2020). https://doi.org/10.1609/aaai.v34i04.6100
9. Drozdzal, M., Vorontsov, E., Chartrand, G., Kadoury, S., Pal, C.: The importance of skip connections in biomedical image segmentation. In: Carneiro, G., et al. (eds.) LABELS/DLMIA -2016. LNCS, vol. 10008, pp. 179–187. Springer, Cham (2016). https://doi.org/10.1007/978-3-319-46976-8_19

Squeeze-and-Excitation Normalization for Automated Delineation of Head and Neck Primary Tumors in Combined PET and CT Images

Andrei Iantsen$^{(\boxtimes)}$ ⓘ, Dimitris Visvikis ⓘ, and Mathieu Hatt ⓘ

LaTIM, INSERM, UMR 1101, University Brest, Brest, France
andrei.iantsen@inserm.fr

Abstract. Development of robust and accurate fully automated methods for medical image segmentation is crucial in clinical practice and radiomics studies. In this work, we contributed an automated approach for Head and Neck (H&N) primary tumor segmentation in combined positron emission tomography/computed tomography (PET/CT) images in the context of the MICCAI 2020 Head and Neck Tumor segmentation challenge (HECKTOR). Our model was designed on the U-Net architecture with residual layers and supplemented with Squeeze-and-Excitation Normalization. The described method achieved competitive results in cross-validation (DSC 0.745, precision 0.760, recall 0.789) performed on different centers, as well as on the test set (DSC 0.759, precision 0.833, recall 0.740) that allowed us to win first prize in the HECKTOR challenge among 21 participating teams. The full implementation based on PyTorch and the trained models are available at https://github.com/iantsen/hecktor.

Keywords: Medical imaging · Segmentation · Head and neck cancer · U-Net SE Normalization

1 Introduction

Combined positron emission tomography/computed tomography (PET/CT) imaging is broadly used in clinical practice for radiotherapy treatment planning, initial staging and response assessment. In radiomics analyses, quantitative evaluation of radiotracer uptake in PET images and tissues density in CT images, aims at extracting clinically relevant features and building diagnostic, prognostic and predictive models. The segmentation step of the radiomics workflow is the most time-consuming bottleneck and variability in usual semi-automatic segmentation methods can significantly affect the extracted features, especially in case of manual segmentation, which is affected by the highest magnitude of inter- and intra-observer variability. Under these circumstances, a fully automated segmentation is highly desirable to automate the whole process and facilitate its clinical routine usage.

V. Andrearczyk et al. (Eds.): HECKTOR 2020, LNCS 12603, pp. 37–43, 2021.
https://doi.org/10.1007/978-3-030-67194-5_4

The MICCAI 2020 Head and Neck Tumor segmentation challenge (HECK-TOR) [1] aims at evaluating automatic algorithms for segmentation of Head and Neck (H&N) tumors in combined PET and CT images. A dataset of 201 patients from four medical centers in Québec (CHGJ, CHMR, CHUM and CHUS) with histologically proven H&N cancer in the oropharynx is provided for a model development. A test set comprised of 53 patients from a different center in Switzerland (CHUV) is used for evaluation. All images were re-annotated by an expert for the purpose of the challenge in order to determine primary gross tumor volumes (GTV) on which the methods are evaluated using the Dice score (DSC), precision and recall.

This paper describes our approach based on convolutional neural networks supplemented with Squeeze-and-Excitation Normalization (SE Normalization or SE Norm) layers to address the goal of the HECKTOR challenge.

2   Materials and Methods

2.1   SE Normalization

The key element of our model is SE Normalization layers [6] that we recently proposed in the context of the Brain Tumor Segmentation Challenge (BraTS 2020) [3]. Similarly to Instance Normalization [4], for an input $X = (x_1, x_2, \ldots, x_N)$ with N channels, SE Norm layer first normalizes all channels of each example in a batch using the mean and standard deviation:

$$x'_i = \frac{1}{\sigma_i}(x_i - \mu_i) \tag{1}$$

where $\mu_i = \mathrm{E}[x_i]$ and $\sigma_i = \sqrt{\mathrm{Var}[x_i] + \epsilon}$ with ϵ as a small constant to prevent division by zero. After, a pair of parameters γ_i, β_i are applied to each channel to scale and shift the normalized values:

$$y_i = \gamma_i x'_i + \beta_i \tag{2}$$

In case of Instance Normalization, both parameters γ_i, β_i, fitted in the course of training, stay fixed and independent on the input X during inference. By contrast, we propose to model the parameters γ_i, β_i as functions of the input X by means of Squeeze-and-Excitation (SE) blocks [5], i.e.

$$\gamma = f_\gamma(X) \tag{3}$$

$$\beta = f_\beta(X) \tag{4}$$

where $\gamma = (\gamma_1, \gamma_2, \ldots, \gamma_N)$ and $\beta = (\beta_1, \beta_2, \ldots, \beta_N)$ - the scale and shift parameters for all channels, f_γ - the original SE block with the sigmoid, and f_β is modeled as the SE block with the tanh activation function to enable the negative shift (see Fig. 1a). Both of SE blocks first apply global average pooling (GAP) to squeeze each channel into a single descriptor. Then, two fully connected (FC) layers aim at capturing non-linear cross-channel dependencies. The first FC layer is implemented with the reduction ratio r to form a bottleneck for controlling model complexity. Throughout this paper, we apply SE Norm layers with the fixed reduction ration $r = 2$.

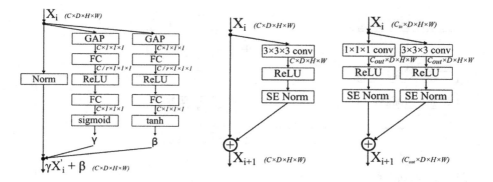

Fig. 1. Layers with SE Normalization: (a) SE Norm layer, (b) residual layer with the shortcut connection, and (c) residual layer with the non-linear projection. Output dimensions are depicted in italics.

2.2 Network Architecture

Our model is built upon a seminal U-Net architecture [7,8] with the use of SE Norm layers [6]. Convolutional blocks, that form the model decoder, are stacks of $3 \times 3 \times 3$ convolutions and ReLU activations followed by SE Norm layers. Residual blocks in the encoder consist of convolutional blocks with shortcut connections (see Fig. 1b). If the number of input/output channels in a residual block is different, a non-linear projection is performed by adding the $1 \times 1 \times 1$ convolutional block to the shortcut in order to match the dimensions (see Fig. 1c).

In the encoder, downsampling is done by applying max pooling with the kernel size of $2 \times 2 \times 2$. To linearly upsample feature maps in the decoder, $3 \times 3 \times 3$ transposed convolutions are used. In addition, we supplement the decoder with three upsampling paths to transfer low-resolution features further in the model by applying the $1 \times 1 \times 1$ convolutional block to reduce the number of channels, and utilizing trilinear interpolation to increase the spatial size of the feature maps (see Fig. 2, yellow blocks).

The first residual block placed after the input is implemented with the kernel size of $7 \times 7 \times 7$ to increase the receptive field of the model without significant computational overhead. The sigmoid function is applied to output probabilities for the target class.

2.3 Data Preprocessing and Sampling

Both PET and CT images were first resampled to a common resolution of $1 \times 1 \times 1$ mm^3 with trilinear interpolation. Each training example was a patch of $144 \times 144 \times 144$ voxels randomly extracted from a whole PET/CT image, whereas validation examples were received from bounding boxes provided by organizers. Training patches were extracted to include the tumor class with the probability of 0.9 to facilitate model training.

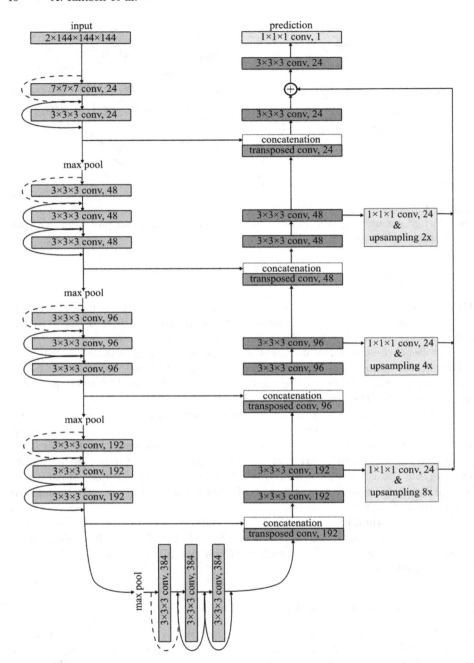

Fig. 2. The model architecture with SE Norm layers. The input consists of PET/CT patches of the size of 144 × 144 × 144 voxels. The encoder consists of residual blocks with identity (solid arrows) and projection (dashed arrows) shortcuts. The decoder is formed by convolutional blocks. Additional upsampling paths are added to transfer low-resolution features further in the decoder. Kernel sizes and numbers of output channels are depicted in each block. (Color figure online)

CT intensities were clipped in the range of $[-1024, 1024]$ Hounsfield Units and then mapped to $[-1, 1]$. PET images were transformed independently with the use of Z-score normalization, performed on each patch.

2.4   Training Procedure

The model was trained for 800 epochs using Adam optimizer on two GPUs NVIDIA GeForce GTX 1080 Ti (11 GB) with a batch size of 2 (one sample per worker). The cosine annealing schedule was applied to reduce the learning rate from 10^{-3} to 10^{-6} within every 25 epochs.

2.5   Loss Function

The unweighted sum of the Soft Dice Loss [9] and the Focal Loss [10] is utilized to train the model. Based on [9], the Soft Dice Loss for one training example can be written as

$$L_{Dice}(y, \hat{y}) = 1 - \frac{2\sum_i^N y_i\hat{y}_i + 1}{\sum_i^N y_i + \sum_i^N \hat{y}_i + 1} \tag{5}$$

The Focal Loss is defined as

$$L_{Focal}(y, \hat{y}) = -\frac{1}{N} \sum_i^N y_i(1 - \hat{y}_i)^\gamma \ln(\hat{y}_i) \tag{6}$$

In both definitions, $y_i \in \{0, 1\}$ - the label for the i-th voxel, $\hat{y}_i \in [0, 1]$ - the predicted probability for the i-th voxel, and N - the total numbers of voxels. Additionally we add $+1$ to the numerator and denominator in the Soft Dice Loss to avoid the zero division in cases when the tumor class is not present in training patches. The parameter γ in the Focal Loss is set at 2.

2.6   Ensembling

Our results on the test set were produced with the use of an ensemble of eight models trained and validated on different splits of the training set. Four models were built using a leave-one-center-out cross-validation, i.e., the data from three centers was used for training and the data from the fourth center was held out for validation. Four other models were fitted on random training/validation splits of the whole dataset. Predictions on the test set were produced by averaging predictions of the individual models and applying a threshold operation with a value equal to 0.5.

Table 1. The performance results on different cross-validation splits. Average results (the row 'Average') are provided for each evaluation metric across all centers in the leave-one-center-out cross-validation (first four rows). The mean and standard deviation of each metric are computed across all data samples in the corresponding validation center. The row 'Average (rs)' indicates the average results on the four random data splits.

Center	DSC	Precision	Recall
CHUS ($n = 72$)	0.744 ± 0.206	0.763 ± 0.248	0.788 ± 0.226
CHUM ($n = 56$)	0.739 ± 0.190	0.748 ± 0.224	0.819 ± 0.216
CHGJ ($n = 55$)	0.801 ± 0.180	0.791 ± 0.208	0.839 ± 0.200
CHMR ($n = 18$)	0.696 ± 0.232	0.739 ± 0.286	0.712 ± 0.228
Average	0.745	0.760	0.789
Average (rs)	0.757	0.762	0.820

Table 2. The test set results of the ensemble of eight models.

Center	DSC	Precision	Recall
CHUV ($n = 53$)	0.759	0.833	0.740

3   Results and Discussion

Our validation results in the context of the HECKTOR challenge are summarized in Table 1. The best outcome in terms of all evaluation metrics was received for the 'CHGJ' center with 55 patients. The model demonstrated the poorest performance for the 'CHMR' center that is least represented in the whole dataset. The differences with the two other centers was minor for all evaluation metrics. The small spread between all centers and the average results implies that the model predictions were robust and any center-specific data standardization was not required. This finding is supported by the lack of significant difference in the average results between the leave-one-center-out and random split cross-validations.

The ensemble results on the test set consisting of 53 patients from the 'CHUV' center are presented in Table 2. On the previously unseen data, the ensemble of eight models achieved the highest results among 21 participating teams with the Dice score of 75.9%, precision 83.3% and recall 74%.

References

1. Andrearczyk, V., et al.: Overview of the HECKTOR challenge at MICCAI 2020: automatic Head and Neck Tumor Segmentation in PET/CT (2021)
2. Andrearczyk, V., et al.: Automatic segmentation of head and neck tumors and nodal metastases in PET-CT scans. In: Medical Imaging with Deep Learning (MIDL) (2020)

3. Menze, B.H., et al.: The multimodal brain tumor image segmentation benchmark (BRATS). IEEE Trans. Med. Imaging **34**(10), 1993–2024 (2015)
4. Ulyanov, D., Vedaldi, A., Lempitsky, V.: Instance normalization: the missing ingredient for fast stylization, arXiv preprint arXiv:1607.08022 (2016)
5. Hu, J., Shen, L., Sun, G.: Squeeze-and-excitation networks, CoRR, vol. abs/1709.01507 (2017). http://arxiv.org/abs/1709.01507
6. Iantsen, A., Jaouen, V., Visvikis, D., Hatt, M.: Squeeze-and-excitation normalization for brain tumor segmentation. In: International MICCAI Brainlesion Workshop (2020)
7. Ronneberger, O., Fischer, P., Brox, T.: U-Net: convolutional networks for biomedical image segmentation. In: Navab, N., Hornegger, J., Wells, W.M., Frangi, A.F. (eds.) MICCAI 2015. LNCS, vol. 9351, pp. 234–241. Springer, Cham (2015). https://doi.org/10.1007/978-3-319-24574-4_28
8. Çiçek, Ö., Abdulkadir, A., Lienkamp, S.S., Brox, T., Ronneberger, O.: 3D U-Net: learning dense volumetric segmentation from sparse annotation. In: Ourselin, S., Joskowicz, L., Sabuncu, M.R., Unal, G., Wells, W. (eds.) MICCAI 2016. LNCS, vol. 9901, pp. 424–432. Springer, Cham (2016). https://doi.org/10.1007/978-3-319-46723-8_49
9. Milletari, F., Navab, N., Ahmadi, S.-A.: V-net: fully convolutional neural networks for volumetric medical image segmentation. In: International Conference on 3D Vision, pp. 565–571. IEEE (2016)
10. Lin, T.-Y., Goyal, P., Girshick, R., He, K., Dollár, P.: Focal Loss for Dense Object Detection, arXiv preprint arXiv:1708.02002 (2017)

Automatic Head and Neck Tumor Segmentation in PET/CT with Scale Attention Network

Yading Yuan[✉]

Department of Radiation Oncology, Icahn School of Medicine at Mount Sinai,
New York, NY, USA
yading.yuan@mssm.edu

Abstract. Automatic segmentation is an essential but challenging step for extracting quantitative imaging bio-markers for characterizing head and neck tumor in tumor detection, diagnosis, prognosis, treatment planning and assessment. The HEad and neCK TumOR Segmentation Challenge 2020 (HECKTOR 2020) provides a common platform for comparing different automatic algorithms for segmentation the primary gross target volume (GTV) in the oropharynx region on FDG-PET and CT images. We participated in the image segmentation challenge by developing a fully automatic segmentation network based on encoder-decoder architecture. In order to better integrate information across different scales, we proposed a dynamic scale attention mechanism that incorporates low-level details with high-level semantics from feature maps at different scales. Our framework was trained using the 201 challenge training cases provided by HECKTOR 2020, and achieved an average Dice Similarity Coefficient (DSC) of 0.7505 with cross validation. By testing on the 53 testing cases, our model achieved an average DSC, precision and recall of 0.7318, 0.7851, and 0.7319 respectively, which ranked our method in the fourth place in the challenge (id: deepX).

1 Introduction

Head and Neck (H&N) cancers are among the most common cancers worldwide (the 5th leading cancer by incidence) [1]. Radiotherapy (RT) combined with chemotherapy is the standard treatment for patients with inoperable H&N cancers [2]. However, studies showed that locoregional failures occur in up to 40% of patients in the first two years after the treatment [3]. In order to identify patients with a worse prognosis before treatment, several radiomic studies have been recently proposed to leverage the massive quantitative features extracted from high-dimensional imaging data acquired during diagnosis and treatment. While these studies showed promising results, their generalization performance needs to be further validated on large patient cohorts. However, the primary tumors and nodal metastases are currently delineated by oncologists by reviewing both PET and CT images simultaneously, which is impractical and error-prone when

© Springer Nature Switzerland AG 2021
V. Andrearczyk et al. (Eds.): HECKTOR 2020, LNCS 12603, pp. 44–52, 2021.
https://doi.org/10.1007/978-3-030-67194-5_5

scaling up to a massive patient population. In addition, radiation oncologists also need to manually delineate the treatment targets and the organs at risk (OARs) when designing a treatment plan for radiotherapy, which is time-consuming and suffers from inter- and intra-operator variations [4]. As a result, automated segmentation methods have been of great demand to assist clinicians for better detection, diagnosis, prognosis, treatment planing as well as assessment of H&N cancers.

The HEad and neCK TumOR segmentation challenge (HECKTOR) [5,6] aims to accelerate the research and development of reliable methods for automatic H&N primary tumor segmentation on oropharyngeal cancers by providing a large PET/CT dataset that includes 201 cases for model training and 53 cases for testing, as an example shown in Fig. 1. For training cases, the ground truth was annotated by multiple radiation oncologists, either directly on the CT of the PET/CT study (31% of the patients) or on a different CT scan dedicated to treatment planning (69%) where the planning CT was registered to the PET/CT scans [7]. While the testing cases were directly annotated on the PET/CT images. The cases were collected from five different institutions where four of them (CHGJ, CHMR, CHUM and CHUS) will be used for training and the remaining one (CHUV) will be used for testing. Each case includes a co-registered PET-CT set as well as the primary Gross Tumor Volume (GTVt) annotation. A bounding box was also provided to enable the segmentation algorithms focus on the volume of interest (VOI) near GTVt [8]. These images were resampled to $1 \times 1 \times 1$ mm isotropic resolution and then cropped to a volume size of $144 \times 144 \times 144$. The evaluation will be based on Dice Similarity Coefficient (DSC), which is computed only within these bounding boxes at the original CT resolution.

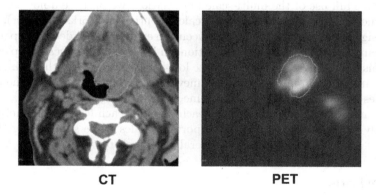

CT   **PET**

Fig. 1. An example of PET/CT used in HECKTOR 2020 challenge

2 Related Work

While convolutional neural networks have been successfully applied in various biomedical image segmentation tasks, only few studies have been conducted in the applications of deep convolutional neural networks (DCNNs) in automated segmentation of tumors in PET/CT images. In [9], Moe et al. presented a PET-CT segmentation algorithm based on 2D U-Net architecture to delineate the primary tumor as well as metastatic lymph nodes. Their model was trained with 152 patients and tested on 40 patients. Andrearczyk et al. [5] expanded this work by investigating several segmentation strategies based on V-Net architecture on a publicly available dataset with 202 patients. Zhao et al. [10] employed a multi-modality fully convolutional network (FCN) [12] for tumor co-segmentation in PET-CT images on a clinic dataset of 84 patients with lung cancer, and Zhong et al. [11] proposed a segmentation method that consists of two coupled 3D U-Nets for simultaneously co-segmenting tumors in PET/CT images for 60 non-small cell lung cancer (NSCLC) patients.

The success of U-Net [13] and its variants in automatic PET-CT segmentation is largely contributed to the skip connection design that allows high resolution features in the encoding pathway be used as additional inputs to the convolutional layers in the decoding pathway, and thus recovers fine details for image segmentation. While intuitive, the current U-Net architecture restricts the feature fusion at the same scale when multiple scale feature maps are available in the encoding pathway. Studies have shown feature maps in different scales usually carry distinctive information in that low-level features represent detailed spatial information while high-level features capture semantic information such as target position, therefore, the full-scale information may not be fully employed with the scale-wise feature fusion in the current U-Net architecture.

To make full use of the multi-scale information, we propose a novel encoder-decoder network architecture named scale attention networks (SA-Net), where we re-design the inter-connections between the encoding and decoding pathways by replacing the scale-wise skip connections in U-Net with full-scale skip connections. This allows SA-Net to incorporate low-level fine details with the high-level semantic information into a unified framework. In order to highlight the important scales, we introduce the attention mechanism [14,15] into SA-Net such that when the model learns, the weight on each scale for each feature channel will be adaptively tuned to emphasize the important scales while suppressing the less important ones. Figure 2 shows the overall architecture of SA-Net.

3 Methods

3.1 Overall Network Structure

SA-Net follows a typical encoding-decoding architecture with an asymmetrically larger encoding pathway to learn representative features and a smaller decoding pathway to recover the segmentation mask in the original resolution. The outputs of encoding blocks at different scales are merged to the scale attention blocks

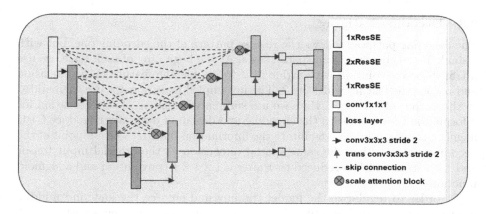

Fig. 2. Architecture of SA-Net. Input is a $2 \times 128 \times 128 \times 128$ tensor followed by one ResSE block with 24 filters. Here ResSE stands for a squeeze-and-excitation block embedded in a residual module [14]. By progressively halving the feature map dimension while doubling the feature width at each scale, the endpoint of the encoding pathway has a dimension of $384 \times 8 \times 8 \times 8$. The output of the encoding pathway has one channel with the same spatial size as the input, i.e., $1 \times 128 \times 128 \times 128$.

(SA-block) to learn and select features with full-scale information. Due to the limit of GPU memory, we convert the input image from $144 \times 144 \times 144$ to $128 \times 128 \times 128$, and concatenate PET and CT images of each patient into a two channel tensor to yield an input to SA-Net with the dimension of $2 \times 128 \times 128 \times 128$. The network output is a map with size of $1 \times 128 \times 128 \times 128$ where each voxel value represents the probability that the corresponding voxel belongs to the tumor target.

3.2 Encoding Pathway

The encoding pathway is built upon ResNet [16] blocks, where each block consists of two Convolution-Normalization-ReLU layers followed by additive identity skip connection. We keep the batch size to 1 in our study to allocate more GPU memory resource to the depth and width of the model, therefore, we use instance normalization, i.e., group normalization [21] with one feature channel in each group, which has been demonstrated with better performance than batch normalization when batch size is small. In order to further improve the representative capability of the model, we add a squeeze-and-excitation module [14] into each residual block with reduction ratio $r = 4$ to form a ResSE block. The initial scale includes one ResSE block with the initial number of features (width) of 24. We then progressively halve the feature map dimension while doubling the feature width using a strided (stride $= 2$) convolution at the first convolution layer of the first ResSE block in the adjacent scale level. All the remaining scales include two ResSE blocks and the endpoint of the encoding pathway has a dimension of $384 \times 8 \times 8 \times 8$.

3.3   Decoding Pathway

The decoding pathway follows the reverse pattern of the encoding one, but with a single ResSE block in each spatial scale. At the beginning of each scale, we use a transpose convolution with stride of 2 to double the feature map dimension and reduce the feature width by 2. The upsampled feature maps are then added to the output of SA-block. Here we use summation instead of concatenation for information fusion between the encoding and decoding pathways to reduce GPU memory consumption and facilitate the information flowing. The endpoint of the decoding pathway has the same spatial dimension as the original input tensor and its feature width is reduced to 1 after a $1 \times 1 \times 1$ convolution and a sigmoid function.

In order to regularize the model training and enforce the low- and middle-level blocks to learn discriminative features, we introduce deep supervision at each intermediate scale level of the decoding pathway. Each deep supervision subnet employs a $1 \times 1 \times 1$ convolution for feature width reduction, followed by a trilinear upsampling layer such that they have the same spatial dimension as the output, then applies a sigmoid function to obtain extra dense predictions. These deep supervision subnets are directly connected to the loss function in order to further improve gradient flow propagation.

3.4   Scale Attention Block

The proposed scale attention block consists of full-scale skip connections from the encoding pathway to the decoding pathway, where each decoding layer incorporates the output feature maps from all the encoding layers to capture fine-grained details and coarse-grained semantics simultaneously in full scales. As an example illustrated in Fig. 3, the first stage of the SA-block is to add the input feature maps at different scales from the encoding pathway, represented as $\{S_e, e = 1, ..., N\}$ where N is the number of total scales in the encoding pathway except the last block ($N = 4$ in this work), after transforming them to the feature maps with the same dimensions, i.e., $S_d = \sum f_{ed}(S_e)$. Here e and d are the scale level at the encoding and decoding pathways, respectively. The transform function $f_{ed}(S_e)$ is determined as follows. If $e < d$, $f_{ed}(S_e)$ downsamples S_e by $2^{(d-e)}$ times by maxpooling followed by a Conv-Norm-ReLU block; if $e = d$, $f_{ed}(S_e) = S_e$; and if $e > d$, $f_{ed}(S_e)$ upsamples S_e through tri-linear upsampling after a Conv-Norm-ReLU block for channel number adjustment. For S_d, a spatial pooling is used to average each feature to form an information embedding tensor $G_d \in R^{C_d}$, where C_d is the number of feature channels in scale d. Then a $1 - to - N$ Squeeze-Excitation is performed in which the global feature embedding G_d is squeezed to a compact feature $g_d \in R^{C_d/r}$ by passing through a fully connected layer with a reduction ratio of r, then another N fully connected layers with sigmoid function are applied for each scale excitation to recalibrate the feature channels on that scale. Finally, the contribution of each scale in each feature channel is normalized with a softmax function, yielding a scale-specific weight vector for each channel as $w_e \in R^{C_d}$, and the final output of the scale attention block is $\widetilde{S}_d = \sum w_e \cdot f_{ed}(S_e)$.

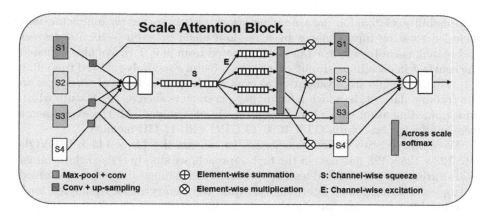

Fig. 3. Scale attention block. Here $S1, S2, S3$ and $S4$ represent the input feature maps at different scales from the encoding pathway.

3.5   Implementation

Our framework was implemented with Python using Pytorch package. All the following steps were performed on the volumes of interest (VOIs) within the given bounding boxes. As for pre-processing, we truncated the CT numbers to $[-125, 225]$ HU to eliminate the irrelevant information, then normalized the CT images with the mean and standard deviation of the HU values within GTVs in the entire training dataset. For PET images, we simply normalized each patient independently by subtracting the mean and dividing by the standard deviation of the image within the body. The model was trained with a patch size of $128 \times 128 \times 128$ voxels and batch size of 1. We used Jaccard distance, which we developed in our previous studies [17–20], as the loss function in this work. It is defined as:

$$L = 1 - \frac{\sum\limits_{i,j,k} (t_{ijk} \cdot p_{ijk}) + \epsilon}{\sum\limits_{i,j,k} (t_{ijk}^2 + p_{ijk}^2 - t_{ijk} \cdot p_{ijk}) + \epsilon}, \tag{1}$$

where $t_{ijk} \in \{0, 1\}$ is the actual class of a voxel x_{ijk} with $t_{ijk} = 1$ for tumor and $t_{ijk} = 0$ for background, and p_{ijk} is the corresponding output from SA-Net. ϵ is used to ensure the stability of numerical computations.

Training the entire network took 300 iterations from scratch using Adam stochastic optimization method. The initial learning rate was set to 0.003, and learning rate decay and early stopping strategies were utilized when validation loss stopped decreasing. In particular, we kept monitoring the validation loss ($L^{(valid)}$) in each iteration. We kept the learning rate unchanged at the first 150 iterations, but dropped the learning rate by a factor of 0.3 when $L^{(valid)}$ stopped improving within the last 30 iterations. The model that yielded the best $L^{(valid)}$ was recorded for model inference.

In order to reduce overfitting, we randomly flipped the input volume in left/right, superior/inferior, and anterior/posterior directions on the fly with a probability of 0.5 for data augmentation. Other geometric augmentations included rotating input images by a random angle between $[-10, 10]$ degrees and scaling them by a factor randomly selected from $[0.9, 1.1]$. We also adjusted the contrast in each image input channel by a factor randomly selected from $[0.9, 1.1]$. We used 5-fold cross validation to evaluate the performance of our model on the training dataset, in which a few hyper-parameters such as the feature width and input dimension were also experimentally determined. All the experiments were conducted on Nvidia GTX 1080 TI GPU with 11 GB memory.

We employed two different strategies to convert the $144 \times 144 \times 144$ VOIs into $128 \times 128 \times 128$ patches. In the first approach, we simply resized the original VOIs during the training and testing phases, in which image data were resampled isotropically using linear interpolation while the binary mark resampled with near neighbor interpolation. In the other approach, we randomly extracted a patch with a size of $128 \times 128 \times 128$ from the VOIs during the training, and applied the sliding window to extract 8 patches from the VOI (2 windows in each dimension) and averaged the model outputs in the overlapping regions before applying a threshold of 0.5 to obtain a binary mask.

4   Results

We trained SA-Net with the training set provided by the HECKTOR 2020 challenge, and evaluated its performance on the training set via 5-fold cross validation. Table 1 shows the segmentation results in terms of DSC for each fold. As compared to the results in Table 2 that were obtained from a model using the standard U-Net skip connections, the proposed SA-Net improved segmentation performance by 3.2% in patching and 1.3% in resizing, respectively.

Table 1. Segmentation results (DSC) of SA-Net in 5-fold cross validation using the training image dataset.

	fold-1	fold-2	fold-3	fold-4	fold-5	Average
Patching	0.772	0.722	0.751	0.759	0.743	0.749
Resizing	0.761	0.730	0.759	0.763	0.745	0.752

Table 2. Segmentation results (DSC) of U-Net in 5-fold cross validation using the training image dataset.

	fold-1	fold-2	fold-3	fold-4	fold-5	Average
Patching	0.731	0.712	0.752	0.704	0.723	0.725
Resizing	0.760	0.719	0.741	0.755	0.734	0.742

When applying the trained models on the 53 challenge testing cases, a bagging-type ensemble strategy was implemented to combine the outputs of these ten models to further improve the segmentation performance, achieving an average of DSC, precision and recall of 0.7318, 0.7851 and 0.7319 respectively, which ranked our method as the fourth place in the challenge.

5 Summary

In this work, we presented a fully automated segmentation model for head and neck tumor segmentation from PET and CT images. Our SA-Net replaces the long-range skip connections between the same scale in the vanilla U-Net with full-scale skip connections in order to make maximum use of feature maps in full scales for accurate segmentation. Attention mechanism is introduced to adaptively adjust the weights of each scale feature to emphasize the important scales while suppressing the less important ones. As compared to the vanilla U-Net structure with scale-wise skip connection and feature concatenation, the proposed scale attention block not only improved the segmentation performance by 2.25%, but also reduced the number of trainable parameters from 17.8M (U-Net) to 16.5M (SA-Net), which allowed it to achieve a top performance with limited GPU resource in this challenge. In addition, the proposed SA-Net can be easily extended to other segmentation tasks. Without bells and whistles, it has achieved the 3rd place in Brain Tumor Segmentation (BraTS) Challenge 2020[1].

Acknowledgment. This work is partially supported by a research grant from Varian Medical Systems (Palo Alto, CA, USA) and grant UL1TR001433 from the National Center for Advancing Translational Sciences, National Institutes of Health, USA.

References

1. Parkin, M., et al.: Global cancer statistics, 2002. CA: Cancer J. Clin. **55**(2), 74–108 (2005)
2. Bonner, J., et al.: Radiotherapy plus cetuximab for localregionally advanced head and neck cancer: 5-year survival data from a phase 3 randomized trial, and relation between cetuximab-induced rash and survival. Lacent Oncol. **11**(1), 21–28 (2010)
3. Chajon, E., et al.: Salivary gland-sparing other than parotid-sparing in definitive head-and-neck intensity-modulated radiotherapy dose not seem to jeopardize local control. Radiat. Oncol. **8**(1), 132 (2013). https://doi.org/10.1186/1748-717X-8-132
4. Gudi, S., et al.: Interobserver variability in the delineation of gross tumor volume and specified organs-at-risk during IMRT for head and neck cancers and the impact of FDG-PET/CT on such variability at the primary site. J. Med. Imaging Radiat. Sci. **48**(2), 184–192 (2017)
5. Andrearczyk, V., et al.: Automatic segmentation of head and neck tumors and nodal metastases in PET-CT scans. In: Proceedings of MIDL 2020, pp. 1–11 (2020)
6. Andrearczyk, V., et al.: Overview of the HECKTOR challenge at MICCAI 2020: automatic head and neck tumor segmentation in PET/CT. In: Andrearczyk, V., et al. (eds.) HECKTOR 2020. LNCS, vol. 12603, pp. 1–21. Springer, Cham (2021)

---

[1] https://www.med.upenn.edu/cbica/brats2020/rankings.html.

7. Vallieres, M., et al.: Radiomics strategies for risk assessment of tumour failure in head-and-neck cancer. Sci. Rep. **7**(1), 10117 (2017)
8. Andrearczyk, V., et al.: Oropharynx detection in PET-CT for tumor segmentation. In: Irish Machine Vision and Image Processing (2020)
9. Moe, Y.M., et al.: Deep learning for automatic tumour segmentation in PET/CT images of patients with head and neck cancers. In: Proceedings of MIDL (2019)
10. Zhao, X., et al.: Tumor co-segmentation in PET/CT using multi-modality fully convolutional neural network. Phys. Med. Biol. **64**, 015011 (2019)
11. Zhong, Z., et al.: Simultaneous cosegmentation of tumors in PET-CT images using deep fully convolutional networks. Med. Phys. **46**(2), 619–633 (2019)
12. Long, J., et al.: Fully convolutional networks for semantic segmentation. In: CVPR, pp. 3431–3440 (2015)
13. Ronneberger, O., Fischer, P., Brox, T.: U-Net: convolutional networks for biomedical image segmentation. In: Navab, N., Hornegger, J., Wells, W.M., Frangi, A.F. (eds.) MICCAI 2015. LNCS, vol. 9351, pp. 234–241. Springer, Cham (2015). https://doi.org/10.1007/978-3-319-24574-4_28
14. Hu, J., et al.: Squeeze-and-excitation networks. In: Proceedings of CVPR 2018, pp. 7132–7141 (2018)
15. Li, X., et al.: Selective kernel networks. In: Proceedings of CVPR 2019, pp. 510–519 (2019)
16. He, K., et al.: Deep residual learning for image recognition. In: Proceedings of CVPR 2016, pp. 770–778 (2016)
17. Yuan, Y., et al.: Automatic skin lesion segmentation using deep fully convolutional networks with Jaccard distance. IEEE Trans. Med. Imaging **36**(9), 1876–1886 (2017)
18. Yuan, Y.: Hierachical convolutional-deconvolutional neural networks for automatic liver and tumor segmentation. arXiv preprint arXiv:1710.04540 (2017)
19. Yuan, Y.: Automatic skin lesion segmentation with fully convolutional-deconvolutional networks. arXiv preprint arXiv:1703.05154 (2017)
20. Yuan, Y., et al.: Improving dermoscopic image segmentation with enhanced convolutional-deconvolutional networks. IEEE J. Biomed. Health Informat. **23**(2), 519–526 (2019)
21. Wu, Y., He, K.: Group normalization. In: Ferrari, V., Hebert, M., Sminchisescu, C., Weiss, Y. (eds.) ECCV 2018. LNCS, vol. 11217, pp. 3–19. Springer, Cham (2018). https://doi.org/10.1007/978-3-030-01261-8_1

Iteratively Refine the Segmentation of Head and Neck Tumor in FDG-PET and CT Images

Huai Chen[1](ID), Haibin Chen[2], and Lisheng Wang[1](✉)(ID)

[1] Department of Automation, Institute of Image Processing and Pattern Recognition, Shanghai Jiao Tong University, Shanghai 200240, People's Republic of China
lswang@sjtu.edu.cn
[2] Perception Vision Medical Technology, Guangzhou, China

Abstract. The automatic segmentation of head and neck (H&N) tumor from FDG-PET and CT images is urgently needed for radiomics. In this paper, we propose a framework to segment H&N tumor automatically by fusing information of PET and CT. In this framework, multiple 3D-Unets are trained one-by-one. The predictions and features of upstream models will be captured as additional information for the next one to further refine the segmentation. Experiments show that iterative refinements can improve the performance. We evaluated our framework on the dataset of HECKTOR2020 (The challenge of head and neck tumor segmentation) and won the 5th place with average DSC, precision and recall of 0.7241, 0.8479, 0.6701 respectively.

Keywords: Head and neck segmentation · PET-CT · HECKTOR2020 · Iterative refinement

1 Introduction

Radiomics, which use quantitative image biomarkers to predict disease characteristics, has shown tremendous potential to optimize patient care. Particularly, radiomics is an important strategy for risk assessment of tumor failure in head and neck (H&N) tumors [13]. For the radiomics analysis of H&N, an expensive and error-prone manual annotation process of regions of interest (ROI) is needed, making the validation on large cohorts hard to implement and the reproducibility poor [2]. Therefore, an algorithm for automatic segmentation of H&N tumors is urgently needed.

To capture excellent algorithm for radiomics, the challenge of HECKTOR2020, which is a part of MICCAI 2020, is organized to identify the best methods to leverage the rich bi-modal information in the context of H&N primary tumor segmentation [1]. In this challenge, both CT images and PET images are provided to obtain the delineation of gross tumor volume.

© Springer Nature Switzerland AG 2021
V. Andrearczyk et al. (Eds.): HECKTOR 2020, LNCS 12603, pp. 53–58, 2021.
https://doi.org/10.1007/978-3-030-67194-5_6

Recently, in medical image analysis, it is widely studied to realize multi-organ and lesion segmentation by combining bi-modal or multi-modal information [9,11,12]. Chen et al. [3] proposed a framework to fuse multiple modalities of MRI (T1, CET1 and T2) to segment nasopharyngeal carcinoma, in which, multiple encoders are set to get modality-specific features from each modality image and a decoder is built to learn comprehensive features and cross-modal inter-dependencies.

For the automatic analysis problem in PET-CT images, several methods have been proposed in different tasks, including lung cancer segmentation [6,7,14] and lesion detection [8]. However, both the low contrast between normal issue and lesions and the big variation of tumor in size and shape make it still a challenge to realize accurate automatic segmentation of tumor. Meanwhile, how to effectively fuse bi-modal information of PET-CT is still needed to be further studied.

To tackle these problems, we propose a deep-learning-based framework to iteratively refine the predictions. This framework is constituted by multiple 3D-Unets [10]. And the embedded features and predictions of upstream Unets will be captured and fed into the next one to make further refinement. We participate the challenge of HECKTOR2020 to evaluate our method and are among the best scoring teams.

2   Method

Our method is a step-by-step refinement framework as illustrated in Fig. 1. The main purpose is to continuously improve the results by iterative refinements. There are total three steps containing step1-segmentation, directly obtaining predictions from original images, step2-segmentation and step3-segmentation, further refining the segmentation by reusing information from upstream models. Specifically speaking, original CT and PET will be firstly concatenated to fuse bi-modal information and fed into the first Unet to get the initial results. Then the results together with the embedded features will be captured to the Unet of step2 to make further processing to get finer predictions. Similarly, step3-segmentation will reuse the results and features of upstream two models.

2.1   Network Architecture

The architecture of our 3D-Unet is shown in Table 1, and it is worth mentioning that the D, H and W are respectively the depth, height and width of features. Encoder is constituted by four encoder blocks, which are formed with 2 convolution layers followed by an instance-normalization layer [4] and a ReLU layer. And after each encoder block, there is a max-pooling layer with kernel of $2 \times 2 \times 2$ and stride of $2 \times 2 \times 2$ to down-sample feature maps. In the decoder part, four decoder blocks are set to fuse high-level and low-level features. The base decoder block is also constructed with 2 convolution layers followed by an instance-normalization and a ReLU. And there is a decovolutional layer to up-sample feature maps before each decoder block.

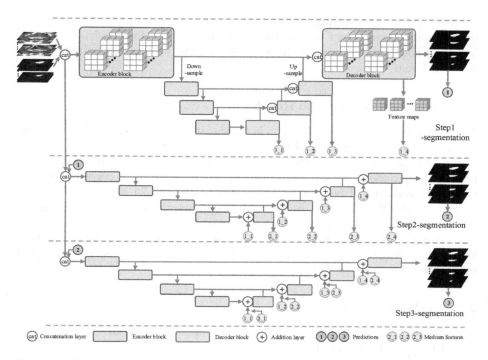

Fig. 1. An illustration of the framework. This method contains three steps: (1) **Step1-segmentation**: Both PET and CT are concatenated and fed into a 3D-Unet containing 4 encoder blocks and 4 decoder blocks. (2) **Step2-segmentation**: Predictions will be further refined based on the results of step1-segmentation, where, the embedded features and outputs of step one are reused in this stage. (3) **Step3-segmentation**: This stage is similar to step two, features and outputs of upstream models are utilized again in this step.

Table 1. The architecture of network.

Stage	Operator	Resolution	Channels	Layers
Encoder part				
1	$Conv3 \times 3 \times 3$	$D \times H \times W$	16	2
2	$Conv3 \times 3 \times 3$	$D/2 \times H/2 \times W/2$	32	2
3	$Conv3 \times 3 \times 3$	$D/4 \times H/4 \times W/4$	64	2
4	$Conv3 \times 3 \times 3$	$D/8 \times H/8 \times W/8$	128	2
Decoder part				
5	$Conv3 \times 3 \times 3$	$D/8 \times H/8 \times W/8$	128	2
6	$Conv3 \times 3 \times 3$	$D/4 \times H/4 \times W/4$	64	2
7	$Conv3 \times 3 \times 3$	$D/2 \times H/2 \times W/2$	32	2
8	$Conv3 \times 3 \times 3$	$D \times H \times W$	16	2
9	$Conv3 \times 3 \times 3$	$D \times H \times W$	1	1

Meanwhile, the corresponding features of encoder and decoder are connected by skip-connection layers to concatenate high-level and low-level information. The final convolution layer is actived by Sigmoid to produce probability values between 0 and 1.

2.2   Iteratively Refine the Results

The purpose of iterative refinement is to repeatedly get finer results based on previous models. As shown in Fig. 1, CT and PET will be firstly concatenated and go through the Unet of step1-segmentation to get initial results. After completing the processing of step one, both embedded features of decoder and segmentation results will be captured as additional inputs for step two. In step2-segmentation, all of CT, PET and results of step one will be concatenated as the input. Meanwhile, the captured embedded features of stage one will be added into corresponding new features to reuse the information. Similarity, in stage three, all of features and results of upstream models will be as the additional information for this new model.

2.3   Training Details

Loss Function: Dice loss is applied as the optimization objective for our model. The definition of Dice loss is shown as followed:

$$Loss_{dice} = -2 \times \frac{\sum_{i=1}^{N} p_i g_i + \epsilon}{\sum_{i=1}^{N} p_i + \sum_{i=1}^{N} g_i + \epsilon} \tag{1}$$

Where ϵ is set as 1 to avoid the risk of being divided by 0. p is denoted as the predicted probability and g is the ground truth.

Training: These three Unets in this framework are trained one by one. Specifically, the Unet of step1-segmentation will be first training with Dice loss. Then, this model will be frozen and the input will go through it to get additional information for step2. Similarly, when training model of step3, both two previous models will be frozen to obtain additional information for new 3D-Unet in this stage.

3   Experiments

Dataset: The challenge of HECKTOR 2020 provides 201 cases from four centers as training data and 53 cases from another center as test data. All of these images are annotated with ground truth of tumor regions by experts, and a bounding box of $144 \times 144 \times 144 \, \text{mm}^3$ is provided for each case to indicate ROI.

Preprocessing: The value of CT and PET are clipped (limited) into $[-250, 250]$ and $[0, 25]$ respectively. Meanwhile, both CT and PET are normalized into $[-1, 1]$ by min-max normalization. For the sampling of both CT and PET, we keep the target spacing of Z-axis as the Z-axis's spacing of original CT, while target spacing of both X-axis and Y-axis are set as 0.97656 mm. The nearest neighbor resampling of SimpleITK is utilized. Additionally, random vertical flipping and random rotation on the XY plane from −15 to 15 are applied.

Implementation Detals: Adam [5] is applied as the optimizer for the training of networks. The learning rate is set as 10^{-3} and the total epochs is 50. Meanwhile, we split 20 cases from the training data as the validation data. With the help of validation set, we can adjust the learning rate to half if the validation loss does not go down in 5 consecutive epochs.

4 Results

4.1 Evaluation Metric

Dice Similarity Coefficient (DSC): The dice similarity coefficient, ranging from 0 to 1, is designed to evaluate the overlap rate of prediction and ground truth. DSC is formulated as followed:

$$DSC(P, G) = \frac{2 \times |P \cap G|}{|P| + |G|} \tag{2}$$

A better result will have a larger DSC.

4.2 Experimental Results

Table 2. Segmentation results on the validation data.

Stage	step1-segmentation	step2-segmentation	step3-segmentation
DSC	0.7476	0.7574	0.7607

The DSC of validation data at different stages are shown in Table 2. Where, DSC is improved by 0.98% and 0.33% respectively in step two and three. We can come to the conclusion that the iterative refinement can continuously improve the performance of tumor segmentation on H&N.

When evaluated on the test data, our method gains the average DSC, precision and recall with 0.7241, 0.8479, 0.6701 respectively.

5 Conclusion

In this paper, we propose a framework to iteratively refine the segmentation of head and neck tumor. In this framework, three segmentation networks are constructed and trained one by one. To realize iterative refinements, the features and results of upstream models will be the additional information for the next one. When evaluated on the validation data, the effectiveness of our framework is proved. And we won the 5th place when evaluated our method on the test data.

References

1. Andrearczyk, V., et al.: Overview of the HECKTOR challenge at MICCAI 2020: automatic head and neck tumor segmentation in PET/CT. In: Andrearczyk, V., et al. (eds.) HECKTOR 2020. LNCS, vol. 12603, pp. 1–21. Springer, Cham (2021)
2. Andrearczyk, V., et al.: Automatic segmentation of head and neck tumors and nodal metastases in PET-CT scans (2020)
3. Chen, H., et al.: MMFNet: a multi-modality MRI fusion network for segmentation of nasopharyngeal carcinoma. Neurocomputing **394**, 27–40 (2020)
4. Huang, X., Belongie, S.: Arbitrary style transfer in real-time with adaptive instance normalization. In: 2017 IEEE International Conference on Computer Vision (ICCV), pp. 1501–1510 (2017)
5. Kingma, D., Ba, J.: Adam: a method for stochastic optimization. international conference on learning representations (2015)
6. Kumar, A., Fulham, M.J., Feng, D.D.F., Kim, J.: Co-learning feature fusion maps from PET-CT images of lung cancer. IEEE Trans. Med. Imaging **39**(1), 204–217 (2019)
7. Li, L., Zhao, X., Lu, W., Tan, S.: Deep learning for variational multimodality tumor segmentation in PET/CT. Neurocomputing **392**, 277–295 (2019)
8. Lina, X., et al.: Automated whole-body bone lesion detection for multiple myeloma on 68ga-pentixafor PET/CT imaging using deep learning methods. Contrast Media Mol. Imaging **2018**, 2391925 (2018)
9. Ma, C., Luo, G., Wang, K.: Concatenated and connected random forests with multiscale patch driven active contour model for automated brain tumor segmentation of MR images. IEEE Trans. Med. Imaging **37**(8), 1943–1954 (2018)
10. Ronneberger, O., Fischer, P., Brox, T.: U-Net: convolutional networks for biomedical image segmentation. In: Navab, N., Hornegger, J., Wells, W.M., Frangi, A.F. (eds.) MICCAI 2015. LNCS, vol. 9351, pp. 234–241. Springer, Cham (2015). https://doi.org/10.1007/978-3-319-24574-4_28
11. Tseng, K.L., Lin, Y.L., Hsu, H.W., Huang, C.Y.: Joint sequence learning and cross-modality convolution for 3D biomedical segmentation. In: 30th IEEE/CVF Conference on Computer Vision and Pattern Recognition (CVPR) (2017)
12. Valindria, V.V., Pawlowski, N., Rajchl, M., Lavdas, I., Glocker, B.: Multi-modal learning from unpaired images: application to multi-organ segmentation in CT and MRI. In: IEEE Winter Conference on Applications of Computer Vision, pp. 547–556 (2018)
13. Vallières, M., et al.: Radiomics strategies for risk assessment of tumour failure in head-and-neck cancer. Sci. Rep. **7**(1), 10117 (2017)
14. Zhao, X., Li, L., Lu, W., Tan, S.: Tumor co-segmentation in PET/CT using multimodality fully convolutional neural network. Phys. Med. Biol. **64**, 015011 (2018)

Combining CNN and Hybrid Active Contours for Head and Neck Tumor Segmentation in CT and PET Images

Jun Ma[1] and Xiaoping Yang[2(✉)]

[1] Department of Mathematics, Nanjing University of Science and Technology, Nanjing, China
junma@njust.edu.cn
[2] Department of Mathematics, Nanjing University, Nanjing, China
xpyang@nju.edu.cn

Abstract. Automatic segmentation of head and neck tumor plays an important role for radiomics analysis. In this short paper, we propose an automatic segmentation method for head and neck tumors from PET and CT images based on the combination of convolutional neural networks (CNNs) and hybrid active contours. Specifically, we first introduce a multi-channel 3D U-Net to segment the tumor with the concatenated PET and CT images. Then, we estimate the segmentation uncertainty by model ensembles, and define a segmentation quality score to select the cases with high uncertainties. Finally, we develop a hybrid active contour model to refine the high uncertainty cases. We evaluate the proposed method on the MICCAI 2020 HECKTOR challenge and achieve promising performance with average Dice Similarity Coefficient, precision and recall of 0.7525, 0.8384, 0.7471 respectively.

Keywords: Segmentation · Deep learning · Uncertainty · Active contours

1 Introduction

Heak and neck cancers are one of the most common cancers [14]. Extracting quantitative image bio-markers from PET and CT images has shown tremendous potential to optimize patient care, such as predicting disease characteristics [9,15]. However, it relies on an expensive and error-prone manual annotation process of Regions of Interest (ROI) to focus the analysis. The fully automatic segmentation methods for head and neck tumors in PET and CT images are highly demanded because they will enable the validation of radiomics models on very large cohorts and with optimal reproducibility.

PET and CT modalities include complementary and synergistic information for tumor segmentation. Thus, the key is how to explore the complementary information. Several methods have been proposed for joint PET and CT segmentation. Kumar et al. [8] proposed a co-learning CNN to improve the fusion

© Springer Nature Switzerland AG 2021
V. Andrearczyk et al. (Eds.): HECKTOR 2020, LNCS 12603, pp. 59–64, 2021.
https://doi.org/10.1007/978-3-030-67194-5_7

of complementary information in multi-modality PET-CT, which includes two modality-specific encoders, a co-learning component, and a reconstruction component. Li et al. [11] proposed a deep learning based variational method for non-small cell lung cancer segmentation. Specifically, A 3D fully convolutional network (FCN) was traind on CT images to produce a probability. Then, A fuzzy variational model was then proposed to incorporate the probability map and the PET intensity image. A split Bregman algorithm was used to minimize the variational model. Recently, Andrearczyk et al. [1] used 2D and 3D V-Net to segment head and neck tumor from PET and CT images. Results showed that using the two modalities can obtain a statistically significant improvement than using CT images or PET images only.

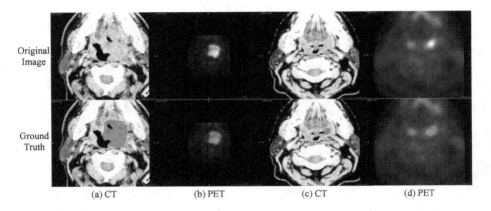

Fig. 1. Visual examples of PET and CT image and the corresponding ground truth.

Active contours [3,4,7] have been one of the widely used segmentation methods before deep learning ear. The basic idea is to formulate the image segmentation task as an energy functional minimization problem. According to used the information in the energy functional, active contours can be classified into three categories, including edge-based active contours [3] that rely on image gradient information, region-based active contours [10] that rely on image-intensity region descriptors, and hybrid active contours [17,18] that use both image gradient and intensity information.

In this short paper, we propose an automatic segmentation method for head and neck tumors from PET and CT images based on the combination of convolutional neural networks (CNNs) and hybrid active contours. Specifically, we first introduce a multi-channel 3D U-Net to segment the tumor with the concatenated PET and CT images. Then, we estimate the segmentation uncertainty by model ensembles, and define a quality score to select the cases with high uncertainties. Finally, we develop a hybrid active contour model to refine the high uncertainty cases.

2   Method

2.1   CNN Backbone

Our network backbone is the typical 3D U-Net [5]. The number of features is 32 in the first block. In each downsampling stage, the number of features is doubled. The implementation is based on nnU-Net [6]. In particular, the network input is configured with a batch size 2. The patch size is $128 \times 128 \times 128$. The optimizer is stochastic gradient descent with an initial learning rate (0.01) and a nesterov momentum (0.99). To avoid overfitting, standard data augmentation techniques are used during training, such as rotation, scaling, adding Gaussian Noise, gamma correction. The loss function is the sum between Dice loss and TopK loss [12]. We train the 3D U-Net with five-fold cross validation. Each fold is trained on a TITAN V100 GPU with 1000 epochs. The training time costs about 4 days.

2.2   Uncertainty Quantification

We train five U-Net models with five-fold cross validation. During testing, we infer the test cases with the trained five models. Thus, each test case has five predictions. Let p_i denote the predictions (Probability) of the $i - th$ model, the final segmentation S can be obtained by

$$S = \frac{1}{5} \sum_{i=1}^{5} p_i. \tag{1}$$

Then, we compute the normalized surface Dice NSD_i between each prediction and the final segmentation. Details and the code are publicly available at http://medicaldecathlon.com/files/Surface_distance_based_measures.ipynb. Finally, the uncertainty of the prediction is estimated by

$$Unc = 1 - \frac{1}{5} \sum_{i=1}^{5} NSD_i. \tag{2}$$

If one case have a uncertainty value over 0.2, it will be selected for the next refinement.

2.3   Refinement with Hybrid Active Contours

This step aims to refine the segmentation results of the cases with high uncertainties by exploiting the complementary information among CT images, PET images, and network probabilities. Basically, CT images can provide edge informations, and PET and network probabilities can provide location or region informations. We propose the following hybrid active contour model

$$E(u) = E_{PET}(u) + E_{CT}(u) + E_{CNN}(u), \tag{3}$$

where

$$E_{PET}(u; f_1, f_2) = \int_\Omega \int_\Omega K(x,y)|I_{PET}(y) - f_1(x)|^2 u d\mathbf{x}$$
$$+ \int_\Omega \int_\Omega K(x,y)|I_{PET}(y) - f_2(x)|^2 (1-u) dx, \quad (4)$$

$$E_{CT}(u) = \sqrt{\frac{\pi}{\tau}} \int_\Omega \sqrt{g_{CT}} u G_\tau * (\sqrt{g_{CT}}(1-u)) d\mathbf{x} \quad (5)$$

and

$$E_{CNN}(u; c_1, c_2) = \int_\Omega (P_{CNN} - c_1)^2 u + (P_{CNN} - c_2)^2 (1-u) d\mathbf{x}. \quad (6)$$

I is the image intensity values, $K(x,y)$ is the Gaussian kernel function, and c_1, c_2 are the average image intensities inside and outside the segmentation contour, respectively. G_τ is the Gaussian kernel, which is defined by

$$G_\tau(x) = \frac{1}{(4\pi\tau)^{3/2}} \exp(-\frac{|\mathbf{x}|}{4\tau}) \quad (7)$$

The hybrid active contour model is solved by thresholding dynamics where the details can be found at [13, 16].

3 Experiments and Results

3.1 Dataset

We use the official HECKTOR dataset [2] to evaluate the proposed method. The training data comprises 201 cases from four centers (CHGJ, CHMR, CHUM and CHUS). The test data comprise 53 cases from another center (CHUV). Each case comprises: CT, PET and GTVt (primary Gross Tumor Volume) in NIfTI format, as well as the bounding box location and patient information. We use the official bounding box to crop all the images. We also resample the images to isotropic resolution 1 mm × 1 mm × 1 mm. Specifically, We use third order spline interpolation and zero order nearest interpolation for the images and labels, respectively. Furthermore, we apply Z-score (mean subtraction and division by standard deviation) to separately normalize each PET and CT image.

3.2 Quantitative and Qualitative Results

Table 1 and Fig. 2 present the quantitative and qualitative results on the testing set, respectively. The proposed method achieved the 2nd place on the official leaderboard, which is also very close to the 1st-place performance. The segmentation results have better precision but inferior recall, indicating that most of the segmentation results are right but some tumors are missed by the method.

Table 1. Quantitative results on the testing set.

Participants	DSC	Precision	Recall	Rank
andrei.iantsen	**0.759**	0.833	**0.740**	1
junma (**Ours**)	0.752	0.838	0.717	2
badger	0.735	0.833	0.702	3
deepX	0.732	0.785	0.732	4
AIView_sjtu	0.724	**0.848**	0.670	5
DCPT	0.705	0.765	0.705	6

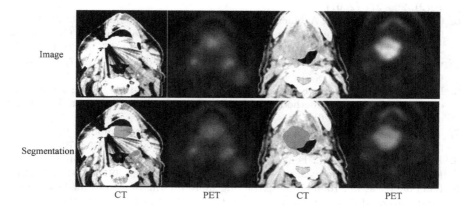

Image

Segmentation

CT   PET   CT   PET

Fig. 2. Visual examples of segmentation results from testing set.

4 Conclusion

In this paper, we proposed a fully automatic segmentation method for head and neck tumor segmentation in CT and PET images, which combines modern deep learning methods and traditional active contours. Experiments on official HECK-TOR challenge dataset demonstrate the effectiveness of the proposed method. The main limitation of our method is the low recall, indicating that some of the lesions are missed in the segmentation results. This would be our further work to enhance the results towards higher performance.

Acknowledgement. This project is supported by the National Natural Science Foundation of China (No. 11531005, No. 11971229). The authors of this paper declare that the segmentation method they implemented for participation in the HECKTOR challenge has not used any pre-trained models nor additional datasets other than those provided by the organizers. We also thanks the HECKTOR organizers for their public dataset and hosting the great challenge.

References

1. Andrearczyk, V., et al.: Automatic segmentation of head and neck tumors and nodal metastases in PET-CT scans. In: Proceedings of Machine Learning Research (2020)
2. Andrearczyk, V., et al.: Overview of the HECKTOR challenge at MICCAI 2020: automatic head and neck tumor segmentation in PET/CT. In: Andrearczyk, V., et al. (eds.) HECKTOR 2020. LNCS, vol. 12603, pp. 1–21. Springer, Cham (2021)
3. Caselles, V., Kimmel, R., Sapiro, G.: Geodesic active contours. Int. J. Comput. Vision **22**(1), 61–79 (1997). https://doi.org/10.1023/A:1007979827043
4. Chan, T.F., Vese, L.A.: Active contours without edges. IEEE Trans. Image Process. **10**(2), 266–277 (2001)
5. Çiçek, Ö., Abdulkadir, A., Lienkamp, S.S., Brox, T., Ronneberger, O.: 3D U-Net: learning dense volumetric segmentation from sparse annotation. In: Ourselin, S., Joskowicz, L., Sabuncu, M.R., Unal, G., Wells, W. (eds.) MICCAI 2016. LNCS, vol. 9901, pp. 424–432. Springer, Cham (2016). https://doi.org/10.1007/978-3-319-46723-8_49
6. Isensee, F., Jäger, P.F., Kohl, S.A., Petersen, J., Maier-Hein, K.H.: Automated design of deep learning methods for biomedical image segmentation. arXiv preprint arXiv:1904.08128 (2020)
7. Kass, M., Witkin, A., Terzopoulos, D.: Snakes: active contour models. Int. J. Comput. Vision **1**(4), 321–331 (1988). https://doi.org/10.1007/BF00133570
8. Kumar, A., Fulham, M., Feng, D., Kim, J.: Co-learning feature fusion maps from PET-CT images of lung cancer. IEEE Trans. Med. Imaging **39**(1), 204–217 (2019)
9. Lambin, P., et al.: Radiomics: extracting more information from medical images using advanced feature analysis. Eur. J. Cancer **48**(4), 441–446 (2012)
10. Li, C., Kao, C.Y., Gore, J.C., Ding, Z.: Minimization of region-scalable fitting energy for image segmentation. IEEE Trans. Image Process. **17**(10), 1940–1949 (2008)
11. Li, L., Zhao, X., Lu, W., Tan, S.: Deep learning for variational multimodality tumor segmentation in PET/CT. Neurocomputing **392**, 277–295 (2020)
12. Ma, J.: Segmentation loss odyssey. arXiv preprint arXiv:2005.13449 (2020)
13. Ma, J., Wang, D., Wang, X.P., Yang, X.: A fast algorithm for geodesic active contours with applications to medical image segmentation. arXiv preprint arXiv:2007.00525 (2020)
14. Siegel, R.L., Miller, K.D., Jemal, A.: Cancer statistics, 2020. CA Cancer J. Clin. **70**(1), 7–30 (2020)
15. Vallieres, M., et al.: Radiomics strategies for risk assessment of tumour failure in head-and-neck cancer. Sci. Rep. **7**(1), 1–14 (2017)
16. Wang, D., Wang, X.P.: The iterative convolution-thresholding method (ICTM) for image segmentation. arXiv preprint arXiv:1904.10917 (2019)
17. Zhang, W., Wang, X., Chen, J., You, W.: A new hybrid level set approach. IEEE Trans. Image Process. **29**, 7032–7044 (2020)
18. Zhang, Y., Matuszewski, B.J., Shark, L.K., Moore, C.J.: Medical image segmentation using new hybrid level-set method. In: 2008 Fifth International Conference Biomedical Visualization: Information Visualization in Medical and Biomedical Informatics, pp. 71–76 (2008)

Oropharyngeal Tumour Segmentation Using Ensemble 3D PET-CT Fusion Networks for the HECKTOR Challenge

Chinmay Rao[1]([✉]), Suraj Pai[1], Ibrahim Hadzic[1], Ivan Zhovannik[1,2],
Dennis Bontempi[1], Andre Dekker[1], Jonas Teuwen[3], and Alberto Traverso[1]

[1] Department of Radiation Oncology (Maastro), GROW School for Oncology,
Maastricht University Medical Centre+, Maastricht, The Netherlands
chinmay.rao@maastro.nl
[2] Department of Radiation Oncology, Radboud Institute of Health Sciences,
Radboud University Medical Centre, Nijmegen, The Netherlands
[3] Department of Medical Imaging, Radboud University Medical Centre, Nijmegen,
The Netherlands

Abstract. Automatic segmentation of tumours and organs at risk can function as a useful support tool in radiotherapy treatment planning as well as for validating radiomics studies on larger cohorts. In this paper, we developed robust automatic segmentation methods for the delineation of gross tumour volumes (GTVs) from planning Computed Tomography (CT) and FDG-Positron Emission Tomography (PET) images of head and neck cancer patients. The data was supplied as part of the MIC-CAI 2020 HECKTOR challenge. We developed two main volumetric approaches: A) an end-to-end volumetric approach and B) a slice-by-slice prediction approach that integrates 3D context around the slice of interest. We exploited differences in the representations provided by these two approaches by ensembling them, obtaining a Dice score of 66.9% on the held out validation set. On an external and independent test set, a final Dice score of 58.7% was achieved.

Keywords: Oropharyngeal cancer · Radiotherapy treatment planning · Automatic segmentation · Multi-modal · PET-CT · 3D U-Net

1 Introduction

According to the European Society for Medical Oncology, Head and Neck Squamous Cell Carcinoma (HNSCC) is the sixth most frequently occurring cancer globally [7]. Among all cancer types, HNSCC accounts for 6% of the occurrences and 1–2% of the deaths. Radiotherapy is standard of care treatment for HNSCC.

C. Rao and S. Pai—Equal contribution.
J. Teuwen and A. Traverso—These authors share senior authorship.

V. Andrearczyk et al. (Eds.): HECKTOR 2020, LNCS 12603, pp. 65–77, 2021.
https://doi.org/10.1007/978-3-030-67194-5_8

PET-CT scans are usually employed for treatment planning. The treatment planning process involves manual contouring of the gross tumour volumes (GTV) which is time-consuming, expensive, and suffers from inter- and intra-reader variability. Accurate and robust automatic segmentation can potentially solve these issues. In addition to treatment planning, the field of radiomics [1] can also benefit from reliable automatic segmentation algorithms for PET-CT images. Radiomics involves predicting tumour characteristics using image-derived quantitative biomarkers. Large scale validation of PET-CT based radiomics models is currently limited by a shortage of PET-CT image datasets containing precise expert-delineated GTVs, and can be tackled by applying automatic segmentation techniques to generate GTV segmentation from unlabelled data. This is the primary motivation behind the inception of the MICCAI 2020: HECKTOR challenge [2,3] which seeks to evaluate bi-modal fusion approaches for segmentation of oropharyngeal GTV in FDG-PET/CT volumes.

To handle the complementary bi-modal information, a variety of approaches have been proposed in recent literature. Andrearczyk et al. [2] employ two simple PET-CT fusion strategies, namely early fusion and late fusion, in a V-Net based framework for segmenting head-and-neck GTV and metastatic lymph nodes. Zhong et al. [18] apply a late fusion approach using two independent 3D U-Nets for PET and CT respectively, and graph-cut co-segmentation to combine their outputs. Novel and specialised deep neural architectures which incorporate fusion of PET and CT-derived information have also been proposed, for example, a two-stream chained architecture [9], a specialised W-Net architecture for bone-lesion detection [16], multi-branched networks that seek to fuse deep features learnt separately from PET and CT and then co-learn the combined features [10,17], and a modular architecture using multi-modal spatial attention [8]. Li et al. [11] propose a hybrid approach that utilises a 3D fully convolutional network to obtain tumour probability map from CT and fuse it with PET data using a fuzzy variational model.

In this paper, we describe an ensemble based segmentation model consisting of two 3D U-Net based networks, and we compare various strategies for combining their outputs based on volumetric Dice score. We explore simple ensembling methods including weighted averaging, union, and intersection operations. We discuss in detail our segmentation approach in Sect. 2.3 and Sect. 2.4. Additionally, we compare commonly used pre-processing methods and investigate the effects of post-processing on the model performance. Details of the pre-processing and post-processing schemes are described in Sect. 2.2 and Sect. 2.5, respectively. The experiments performed for the aforementioned comparison studies as well as their results are documented in Sect. 3. Finally, we make the code for most of the data operations performed publicly available.[1]

---

[1] https://gitlab.com/UM-CDS/projects/image-standardization-and-domain-adaptation/hecktor-segmentation-challenge.

2   Methodology

2.1   Dataset

For training and validating our models, we used the benchmark dataset supplied for the HECKTOR challenge. The training set consists of FDG-PET/CT data and the corresponding GTV segmentation masks from 201 patients diagnosed with oropharyngeal cancer obtained from four centres in Québec (Canada). This corpus of data is a subset of a larger dataset originally proposed by Vallières et al. [15], which is publicly available on The Cancer Imaging Archive [6,12]. For the purpose of the challenge, this subset underwent quality control, including the conversion of raw PET intensties to SUV and the reannotation of the primary GTV for each patient. The test dataset, provided for the HECKTOR challenge and used for the final evaluation of the submissions, is a set of FDG-PET/CT scans from 53 patients from the Centre Hospitalier Universitaire Vaudois (Lausanne, Switzerland). The supplied imaging data vary in physical size, array size and voxel spacing across patients. Hence, for the purpose of standardisation, we cropped all the images to $144 \times 144 \times 144 \, \text{mm}^3$ physical size using simple PET-based brain segmentation. Subsequently, we resampled the scans using 3rd order spline interpolation to have a pixel spacing of $1 \times 1 \, \text{mm}^2$ in the x-y plane and 3 mm spacing between axial slices. These dimensions were chosen by obtaining a distribution of pixel spacing across the entire dataset and choosing the mode of the distribution to minimise oversampling in comparison to isotropic resampling. The supplied HECKTOR training set was randomly split to produce two subsets with 180 and 21 patients for model training and validation respectively. The aforementioned data preparation steps were implemented using code obtained from the public Github repository released by the challenge organisers[2].

2.2   Pre-processing

The voxel intensities of the resampled CT and PET modalities are measured in Hounsfield Units (HU) and Standardised Uptake Values (SUV), respectively. In the supplied dataset, the PET scan intensities were already converted from absolute activity concentration (Bq/mL) and counts (CNTS) units to SUV. We processed the HU values by applying a window between the range $[-150, 150]$ to focus on tissues within the particular range, which include the GTV. We subsequently normalised this to a range of $[0, 1]$. The maximal SUV values are more dynamic in range compared to the HU values, although between a range of $[0, 5]$ the values follow similar distributions across the dataset. This behaviour of similar distributions between $[0, 5]$ can be seen in Fig. 1. In order to account for this, we limited the SUV values between the range $[0.01, 8]$ following which a $[0, 1]$ normalisation was performed. We refrained from using global normalisation schemes to avoid value shift on the test data which was collected from a different centre. As an alternate normalisation scheme, z-score normalisation was

---

[2] http://github.com/voreille/hecktor

also explored for the PET-CT pair but min-max [0, 1] normalisation ultimately provided the best performance on our validation data.

2.3   Network Architectures

In this study, we test two different network architectures, comparing the strengths and drawbacks of their associated input-output representations. The first network architecture infers the segmentation masks on a slice-by-slice basis, by integrating "a slice context" around each slice to be predicted. At each of these slices, the network outputs predictions based on values of the slice and the values of its neighbours in a certain range. This range of neighbours around a particular slice is what we term as slice context. The second network is a fully volumetric 3D network that takes a full volume as input and outputs another volume containing predictions for each voxel of the input.

3D-to-2D U-Net with Fully Connected Bottleneck. We used a custom implementation of the 3D U-Net network architecture proposed by Nikolov et al. [13]. The input to this network is a 3D volume with 21 slices with a dimension of 128 × 128 each. Of the 21 slices, 10 slices at each side of the central slice comprise the slice context, and the network outputs a 2D segmentation corresponding to the central slice. In terms of network design, 7 down-convolutional blocks with a mix of 2D and 3D convolutions are present in the analysis path of the U-Net. At the end of the analysis path, a fully connected bottleneck is introduced. Following the bottleneck, 7 up-convolutional blocks give way to the synthesis path of the network.

Fully Volumetric 3D U-Net. The 3D U-Net was introduced by Çiçek et al. [5] to extend the success of 2D U-Nets to 3D volumetric inputs. In our work, we used a modified version of the 3D U-Net provided as part of the ELEKTRONN3 Toolkit[3]. The input to this network is a 3D volume with 48 slices in the z axis and each slice has a shape of 144 × 144. The output of the network follows the same spatial configuration as the input. A shallow architecture is used in order to allow fitting more 3D volumes per batch. 3 down-convolutional blocks are used in the analysis path and give way to 2 up-convolutional blocks in the synthesis path.

2.4   Model Training and Hyperparameters

A large amount of focus in our work was placed on model training procedures to account for 3D data, data imbalance, and memory and computational efficiency. For the network described in Sect. 2.3, we used label-based sampling where slices were selected by sampling randomly from all the slices that contain the GTV. In order to account for slices where the GTV is absent we also randomly sampled

---

[3] http://elektronn3.readthedocs.io.

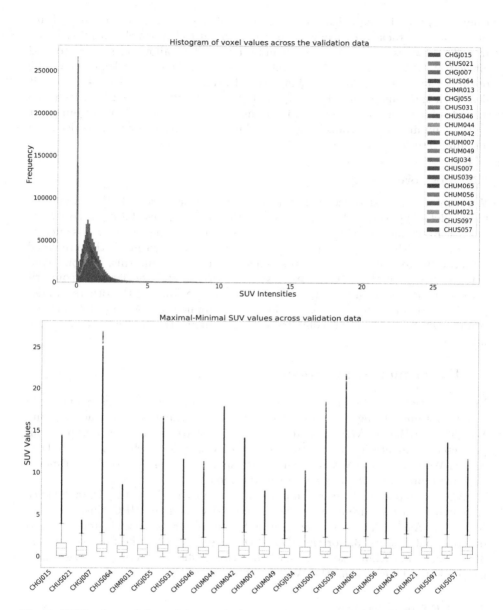

Fig. 1. SUV values of the validation data with a histogram and box plot to show the distribution of voxel intensities and the maximal and minimal values. The legend in figure (a) and the x-axes labels in (b) correspond to subject IDs in the validation data. The first 4 alphabets correspond to a centre and the 3 digits following correspond to a numeric ID. For example, CHUM007 corresponds to a subject 7 from Centre Hospitalier de l'Université de Montréal

from background slices with a certain probability (chosen to be $p = 0.2$). Data imbalance was tackled by using a top-k loss similar to [13] which optimises losses from $k\%$ of worst performing voxels in an image. This allows the model to deal efficiently with imbalanced data points and to train the hardest losses first. The optimiser was attached to a decaying cyclic learning rate scheduler [14] to help it deal with the complex loss landscape obtained through the top-k loss. By using a range of learning rates, the scheduler ensures that the optimiser can jump out of local minima and avoids stagnation during training compared to decaying learning rate schedulers.

2.5   Post-processing

We employed a post-processing step to refine the predicted GTV structure in the model's hypothesised binary masks as well as to address false positive voxel groups. First, morphological dilation was performed using a $5 \times 5 \times 5$ structuring matrix with a roughly spherical structure in order to make the predicted structure more globular as tumours generally are. Following this, all connected components from the binary image were extracted and the largest geometrical structure was considered the GTV while disregarding all the others as false positives. Finally, a morphological closing was applied on the largest connected component to smooth the contours with a similar structuring matrix as the dilation.

3   Experiments and Results

In this section, we describe the experiments performed and consequently the results obtained using our methodologies. All experiments were tracked using Weights and Biases (W&B) [4] to observe qualitative metrics such as per scan predicted segmentation maps and quantitative metrics such as loss and Dice scores. We provide the W&B run info corresponding to each of our experiments to allow reproducibility, hosted on this dashboard.

All our experiments were run on clusters provided by the Data Science Research Infrastructure at Maastricht University[4] and the HPC cluster hosted by RWTH-Aachen[5]. Due to differences in hardware across these clusters, we used different batch sizes (32 for the 3D-to-2D U-Net and 8 for the fully volumetric 3D U-Net) and caching methodologies to perform efficient training. These details can be found by exploring the W&B dashboard.

3.1   PET only Training

As a preliminary experiment, the network in Sect. 2.3 was trained only on PET data. After training for 200,000 iterations[6] using the training configurations mentioned in Sect. 2.4, with a learning rate range of 0.001 to 0.01, a final Dice score

---

[4] https://maastrichtu-ids.github.io/dsri-documentation/.

[5] https://doc.itc.rwth-aachen.de/.

[6] Each iteration corresponds to one forward-backward pass over a batch.

of 0.526 was obtained on the held out validation data. Qualitatively inspecting the obtained results showed that there were numerous cases where high probability of GTV was seen when the PET intensity was high but there was no tumour present in the ground truth. Figure 2 shows an example of the false positives seen. Pairing this PET data with structural information would play a strong role in avoiding such cases and discriminating between false positives in high intensity regions.

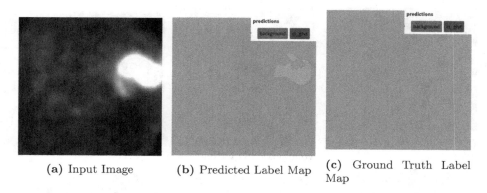

(a) Input Image   (b) Predicted Label Map   (c) Ground Truth Label Map

Fig. 2. False positives predicted by the PET-only network when high intensity regions are seen in the validation set. The legend of semantic labels can be seen on to the top right corner of the label map images. *ct_gtvt* label map corresponds to the presence of tumour within that region while *background* corresponds to its absence.

3.2   PET-CT Early Fusion

To fuse information from both the PET and CT modalities, we applied a very straightforward channel-wise fusion strategy. The PET and CT 3D volumes were stacked across channels forming a 4D PET-CT input to the models. We followed this fusion strategy to allow the entire model to have access to combined PET-CT information as they are complementary in determining GTV contours.

The PET-CT data was fed as input into both the networks defined in Sect. 2.3. For the 3D-to-2D model described in Sect. 2.3, we used a large batch size owing to the smaller 3D input in comparison to the end-to-end volumetric 3D approach. Both networks were also run for 200,000 iterations with these batch sizes. The 3D-to-2D network was trained with rotate, shear and elastic deformations applied on the fly during training time. The results of the training led to qualitatively and quantitatively superior results compared to the PET-only network achieving a Dice score of 0.648 on the validation data split. Qualitatively the results also show increased true positives and a huge decrease in false positives that spiked with higher intensity values as seen in the PET-only approach. A fully volumetric 3D approach was also experimented with—to compare against the 3D volume to 2D slice prediction input-output representation. This experiment provided quantitative results similar to the former approach with a Dice score of 0.639 but differed in the qualitative predictions. The qualitative predictions of this network were significantly smoother in the 3D space than the previous approach but some smaller contours (occupying smaller dimensions in the voxel space) were missed. Figure 3 shows these qualitative differences across different networks.

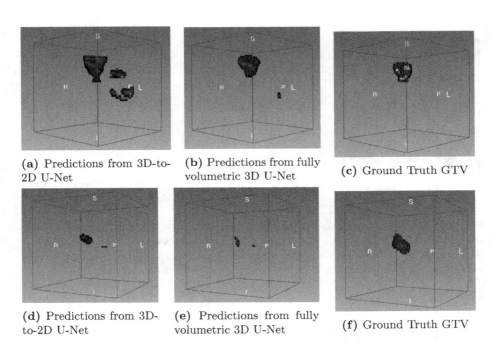

(a) Predictions from 3D-to-2D U-Net

(b) Predictions from fully volumetric 3D U-Net

(c) Ground Truth GTV

(d) Predictions from 3D-to-2D U-Net

(e) Predictions from fully volumetric 3D U-Net

(f) Ground Truth GTV

Fig. 3. GTV predictions for the two types of models compared with the ground truth in 3D. For the images in the first row, (b) has significantly fewer false positives compared to (a) and produces more accurate 3D volumes. In the second row, (d) trumps (e) in terms of correspondence to the ground truth. (e) misses out largely in matching the shape of the contour in (f)

3.3   Model Ensembling

From visual inspection of a subset of predictions from both 3D-to-2D and fully volumetric 3D early fusion U-Nets, we found that each of these networks failed in ways different from the other in correctly segmenting the GTV. This can be seen in Fig. 3. In order to utilise this apparent complementary behaviour, we experimented combining their outputs using simple ensembling approaches to produce the final segmentation mask for each of the validation examples. In particular, we compared three operations - weighted voxel-wise average, union and intersection. The weighted average operation was applied to the output voxel-wise probabilities of the two networks where a single fixed weight was assigned to the output probability map of each network. Union and intersection operations were applied to the binary mask outputs of the two networks each obtained using a GTV probability threshold of 0.5.

3.4   Post-processing

To measure the effect of the post-processing sequence discussed in Sect. 2.5 on the model performance, the post-processing operations were applied to the binary prediction masks of models in every case - the two early-fusion U-Net models and their ensembles - and the resulting validation Dice scores were compared with those of the corresponding models without the post-processing step.

Weight values used for the weighted average ensembling operation were 0.6 and 0.4 for 3D-to-2D and fully volumetric 3D U-Nets respectively when no post-processing was performed. With post-processing, the weights were 0.5 for both. In each case, the weights were optimal among a fixed set of values with respect to the average validation Dice, as shown in Fig. 4.

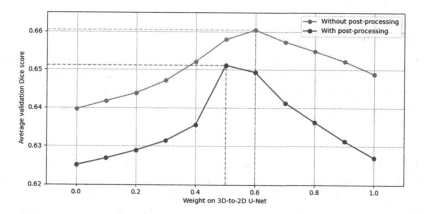

Fig. 4. Weight vs. average validation Dice plot for ensemble model with weighted voxel-wise average strategy with and without the post-processing step. Weight, here, refers to the weight value assigned to the predicted GTV probability map of 3D-to-2D U-Net.

Table 1 shows the validation Dice score for each of the aforementioned model configurations we experimented with. Combining the predictions of the two early fusion U-Nets with weighted voxel-wise average improved the overall performance. This improvement was also observed when post-processing was introduced. The use of post-processing step, however, deteriorated the final score in every case rather than improving it, except for the intersection based ensemble which exhibited a slight improvement. The negative effect of post-processing can be observed on the union ensemble's performance as it sharply dropped from being the best among the others to being the worst. Moreover, in the case of weighted average ensemble, post-processing results in a deteriorated performance for all weights as seen in Fig. 4.

Table 1. Average validation Dice score (DSC) for each of the model and ensemble variants tested.

Network variants	DSC w/o Post-processing	DSC with Post-processing
3D-to-2D U-Net	0.648	0.626
Fully volumetric 3D U-Net	0.639	0.625
Intersection	0.614	0.635
Union	0.669	0.618
Weighted average	0.657	0.651

A block diagram overview of the different components in the segmentation pipeline can be found in Fig. 5. Individual components of the pipeline were described in detail in the preceding sections.

3.5   Post-challenge Results

The best performing model variant on the held out test-set from the challenge was the weighted average ensemble without post-processing applied. This is seen in Fig. 4 at the peak of the green line plot. The 3D-to-2D U-Net predictions, p_1 and the fully volumetric 3D U-Net predictions, p_2 are combined as,

$$p = 0.6 \times p_1 + 0.4 \times p_2 \tag{1}$$

The final binary label map, obtained by thresholding $p >= 0.5$, is submitted to the challenge. With this, we obtain a Dice score of 0.587. Compared to the challenge winners, we see a large drop in our Dice scores (-0.17). We hypothesise this difference to be due to distribution-shift across data from different centres and our method's inability to account for these in the data pre-processing strategies.

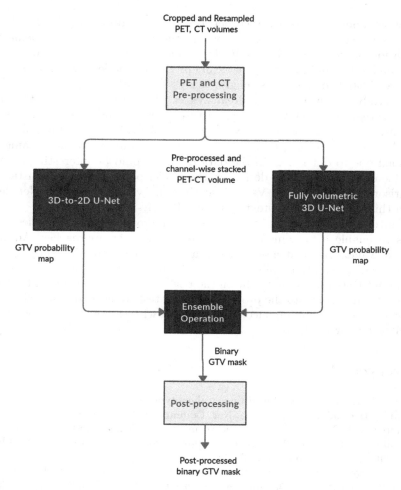

Fig. 5. Block diagram of the segmentation pipeline. The entire procedure followed in obtaining a predicted GTV mask from the PET-CT dataset provided as a part of the challenge is presented in the diagram.

4   Discussion and Conclusion

The HECKTOR challenge provides a strong benchmark to compare automatic segmentation methods for oropharyngeal tumours in PET-CT images. Development of automatic methods can prove to be highly useful in providing delineation assistance in radiotherapy treatment planning as well as for advancement of PET-CT based radiomics by facilitating the generation of segmentation data for validation of radiomics methods on large cohorts. Through the challenge, we were able to compare different 3D approaches with varied input-output representations, pre-processing methods and training hyperparameter schemes.

A stark difference was observed between PET-CT fusion and PET-only network results which quantitatively bolsters the importance of complementary information provided by the combined modalities. After obtaining slice-by-slice prediction models and fully volumetric 3D prediction models, ensemble methods were investigated to combine strengths across these methods.

The study performed by Andrearczyk et al. [2] includes using the early-fusion strategy in a 3D V-Net architecture, among other design choices and combinations, to segment primary orophayngeal GTV and metastatic lymph lodes. Although a meaningful comparison of our results with theirs cannot be performed due to differences in the data used, it would be interesting to study the influence of architecture design on the model performance. For instance, a comparison between the 3D V-Net and the fully volumetric 3D U-Net design used in this study, in the context of PET-CT early-fusion.

As future work, we plan to conduct larger cross validation studies across centres to enable a meaningful comparison with other approaches. Additionally, cross validation strategies that can account for distribution-shift can help us improve generalisation ability of our methods to new test centres as seen in the held out test-set. To shed light on the results in a qualitative manner and to incorporate clinicians into the process, a Turing test could be performed to analyse how satisfied a radiation oncologist would be with the tumours automatically delineated by our methods.

References

1. Aerts, H.J., et al.: Decoding tumour phenotype by noninvasive imaging using a quantitative radiomics approach. Nat. Commun. **5**(1), 1–9 (2014)
2. Andrearczyk, V., et al.: Automatic segmentation of head and neck tumors and nodal metastases in PET-CT scans. In: International Conference on Medical Imaging with Deep Learning (MIDL) (2020)
3. Andrearczyk, V., et al.: Overview of the HECKTOR challenge at MICCAI 2020: automatic head and neck tumor segmentation in PET/CT. In: Andrearczyk, V., et al. (eds.) HECKTOR 2020. LNCS, vol. 12603, pp. 1–21. Springer, Cham (2021)
4. Biewald, L.: Experiment tracking with weights and biases (2020). https://www.wandb.com/. Software available from wandb.com
5. Çiçek, Ö., Abdulkadir, A., Lienkamp, S.S., Brox, T., Ronneberger, O.: 3D U-Net: learning dense volumetric segmentation from sparse annotation. In: Ourselin, S., Joskowicz, L., Sabuncu, M.R., Unal, G., Wells, W. (eds.) MICCAI 2016. LNCS, vol. 9901, pp. 424–432. Springer, Cham (2016). https://doi.org/10.1007/978-3-319-46723-8_49
6. Clark, K., et al.: The cancer imaging archive (TCIA): maintaining and operating a public information repository. J. Digit. Imaging **26**(6), 1045–1057 (2013). https://doi.org/10.1007/s10278-013-9622-7
7. Economopoulou, P., Psyrri, A.: Head and Neck Cancers: Essentials for Clinicians, chap. 1. ESMO Educational Publications Working Group (2017)
8. Fu, X., Bi, L., Kumar, A., Fulham, M., Kim, J.: Multimodal spatial attention module for targeting multimodal PET-CT lung tumor segmentation. arXiv preprint arXiv:2007.14728 (2020)

9. Jin, D., et al.: Accurate esophageal gross tumor volume segmentation in PET/CT using two-stream chained 3D deep network fusion. In: Shen, D., et al. (eds.) MIC-CAI 2019. LNCS, vol. 11765, pp. 182–191. Springer, Cham (2019). https://doi.org/10.1007/978-3-030-32245-8_21

10. Kumar, A., Fulham, M., Feng, D., Kim, J.: Co-learning feature fusion maps from PET-CT images of lung cancer. IEEE Trans. Med. Imaging 39(1), 204–217 (2019)

11. Li, L., Zhao, X., Lu, W., Tan, S.: Deep learning for variational multimodality tumor segmentation in PET/CT. Neurocomputing 392, 277–295 (2020)

12. Martin, V., et al.: Data from head-neck-PET-CT. The Cancer Imaging Archive (2017)

13. Nikolov, S., et al.: Deep learning to achieve clinically applicable segmentation of head and neck anatomy for radiotherapy. CoRR abs/1809.04430 (2018). http://arxiv.org/abs/1809.04430

14. Smith, L.N.: No more pesky learning rate guessing games. CoRR abs/1506.01186 (2015). http://arxiv.org/abs/1506.01186

15. Vallieres, M., et al.: Radiomics strategies for risk assessment of tumour failure in head-and-neck cancer. Sci. Rep. 7(1), 1–14 (2017)

16. Xu, L., et al.: Automated whole-body bone lesion detection for multiple myeloma on 68ga-pentixafor PET/CT imaging using deep learning methods. Contrast Media Mol. Imaging 2018 (2018)

17. Zhao, X., Li, L., Lu, W., Tan, S.: Tumor co-segmentation in PET/CT using multi-modality fully convolutional neural network. Phys. Med. Biol. 64(1), 015011 (2018)

18. Zhong, Z., et al.: 3D fully convolutional networks for co-segmentation of tumors on PET-CT images. In: 2018 IEEE 15th International Symposium on Biomedical Imaging (ISBI 2018), pp. 228–231. IEEE (2018)

Patch-Based 3D UNet for Head and Neck Tumor Segmentation with an Ensemble of Conventional and Dilated Convolutions

Kanchan Ghimire[1(✉)], Quan Chen[1,2], and Xue Feng[1,3]

[1] Carina Medical, Lexington, KY 40513, USA
kghimire@carinaai.com
[2] Department of Radiation Medicine, University of Kentucky,
Lexington, KY 40535, USA
[3] Department of Biomedical Engineering, University of Virginia,
Charlottesville, VA 22903, USA
xf4j@virginia.edu

Abstract. Automatic segmentation of tumor eliminates problems associated with manual annotation of region-of-interest (ROI) from medical images, such as significant human efforts and inter-observer variability. Accurate segmentation of head and neck tumor has a tremendous potential for better radiation treatment planning for cancer (such as oropharyngeal cancer) and also for optimized patient care. In recent times, the development in deep learning models has been able to effectively and accurately perform segmentation tasks in semantic segmentation as well as in medical image segmentation. In medical imaging, different modalities focus on different properties and combining the information from them can improve the segmentation task. In this paper we developed a patch-based deep learning model to tackle the memory issue associated with training the network on 3D images. Furthermore, an ensemble of conventional and dilated convolutions was used to take advantage of both methods: the smaller receptive field of conventional convolution allows to capture finer details, whereas the larger receptive field of dilated convolution allows to capture better global information. Using patch-based 3D UNet with an ensemble of conventional and dilated convolution yield promising result, with a final dice score of 0.6911.

Keywords: Head and neck tumor segmentation · Ensemble · Patch-based segmentation · 3D UNet · Deep learning

1 Introduction

Oropharyngeal cancer is a disease in which malignant, or cancer, cells form in the tissues of oropharynx. Oropharynx, comprising of tongue base, soft palate, tonsils and side and back walls of the throat, is one of the most common sites for head and neck cancer. Accurate segmentation of head and neck tumor from medical images could result in better radiation treatment planning and optimized patient care based on the identification of exact tumor location. The development of deep learning models in

© Springer Nature Switzerland AG 2021
V. Andrearczyk et al. (Eds.): HECKTOR 2020, LNCS 12603, pp. 78–84, 2021.
https://doi.org/10.1007/978-3-030-67194-5_9

recent times have been able to effectively and accurately perform medical image segmentation task, while eliminating the problems associated with manual annotation of region-of-interest (ROI) such as significant human efforts and inter-observer variability.

To compare and evaluate different automatic-segmentation algorithms, Medical Image Computing and Computer Assisted Intervention (MICCAI) 2020 Challenge was organized using FDG-PET and CT scans of head and neck (H&N) primary tumors. More specifically, the dataset used in this challenge includes multiple-institutional clinically acquired bimodal scans (FDG-PET and CT), focusing on oropharyngeal cancers. Since training deep learning model on 3D images pose memory issue, patch-based approach was used in this study. Furthermore, an ensemble of conventional and dilated convolution network was performed. The smaller receptive field of conventional convolution allows to capture fine details and the larger receptive field of dilated convolution allows to capture better global information, and the ensemble method allows to take advantage of both. Similarly, ensemble modeling technique also decreases the generalization error and helps to improve the predictive performance when compared to individual models.

2 Methods

For the automatic segmentation of head and neck tumor task, the steps in our proposed method include image pre-processing, patch extraction, training multiple models using a generic 3D U-Net structure with different hyper-parameters, deployment of each model for prediction, final ensemble modeling and image post-processing. Details are described in different sections as follows.

2.1 Image Pre-processing

Positron Emission Tomography (PET) measures metabolic activity of the cells of the body tissues. Computed Tomography, on the other hand, focuses on the morphological tissue properties. Fusing the information from the two modalities, PET and CT, was one of the major tasks during image pre-processing. CT scans are expressed in Hounsfield units (HU), and since CT scan images in our dataset include head and neck region, -1024 HU to 1024 HU was selected as an appropriate window. Each 3D CT image was normalized to 0 to 1 by dividing the HU values by the window range. Since bounding boxes were provided in the dataset, each 3D CT images were cropped using bounding boxes coordinates provided in patient reference. To do so, bounding boxes coordinates were first translated into their respective index values. Similarly, each 3D PET image was cropped using bounding boxes and resampled to the size of their respective 3D CT images using linear interpolation method. Then, for each subject, 3D CT and 3D PET images were fused along the last dimension so that the input image has two channels. As FDG-PET images do not have standard pixel intensity values, to reduce the effects from different contrasts and different subjects, each 3D image was normalized by subtracting their global mean values and divided by the pixel intensity global standard deviation values.

2.2 Patch Extraction

In a 3D U-Net model architecture, using large input size or the entire image poses several challenges. Some of the challenges include memory issue, long training time and class imbalance. Since the memory of a moderate GPU is about 11 Gb or 12 Gb, the network needs to greatly reduce the number of features and/or the layers to fit the model into the GPU. Reducing the number of features and/or the layers, oftentimes, leads to reduced network performance. Similarly, the training time will increase significantly as more voxels contribute to calculation of the gradients at each step and the number of steps cannot be proportionally reduced during optimization. Since the portion of the image with tumor is usually relatively smaller compared to the whole image, there will be a class imbalance problem. The class imbalance, therefore, will cause the model to focus on non-tumor voxels if trained with uniform loss, or will be prone to false positives if trained with weighted loss that favors the tumor voxels. Therefore, to more effectively utilize the training data, patch-based segmentation approach was applied where smaller patches were extracted from each subject. Patch-based approach also provides a natural way to do data augmentation since we can extract patches differently. The patch size was as a tunable hyper parameter during training. While larger patch size helps extract more global information from medical images, it also comes with the cost of memory issue and longer training time. For implementation, a random patch was extracted from each subject using uniform probabilities during each epoch. The data augmentation applied was the random left-right flip after patch extraction. A random left-right flip was performed since the head and neck region is anatomically symmetric along left-right direction.

2.3 Network Structure and Training

A 3D U-Net network structure with 3 encoding and 3 decoding blocks was used, as shown in Fig. 1. For each encoding block, a VGG like network with two consecutive 3D convolutional layers was used with kernel size of 3. It was followed by the rectified linear unit (ReLU) activation function. Finally, batch normalization layers were used. A large number of features were used in the first encoding block to improve the expressiveness of the network. To combat overfitting, a dropout ratio of 0.5 was added after the last encoding block. Weighted cross entropy and dice was used as the loss function, with 1.0 value used for non-tumor voxels and 2.0 value used for tumor voxels. Similar to convention U-Net structure, the spatial dimension was halved whereas the number of features was doubled during each encoding block. For the decoding blocks, symmetric blocks were used with skip-connections from corresponding encoding blocks. Features were concatenated to the de-convolution outputs, and the segmentation map of the input patch was expanded to the binary class (tumor and non-tumor/background) ground truth labels.

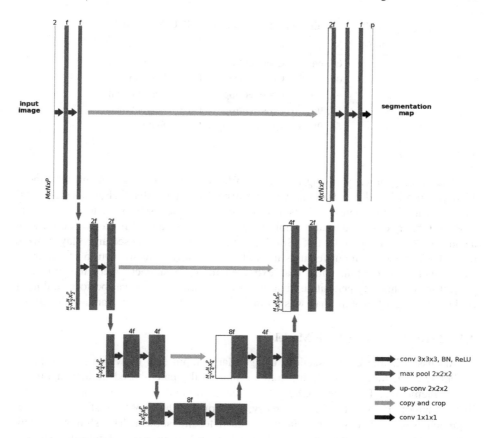

Fig. 1. 3D U-Net structure with 3 encoding and 3 decoding blocks.

Hyperparameter tuning was performed for model selection, where multiple models were created using tunable parameters: 1) patch size; 2) loss type; 3) type of convolution. Cross-validation was applied during hyperparameter tuning, where 60% of the training data was used as training set and the rest 40% was used as validation set. While larger patch size could significantly increase the training time, it also has the potential to significantly improve the model performance. Similarly, using dilated convolution allows the network to have a larger receptive field and could lead to extracting better global information. Table 1. shows the parameters used to train five different models, where N denotes the input size, Loss Type denotes the loss used and Convolution denotes whether conventional convolution or dilated convolution was used in the network. During model selection process, both dice score and loss value was used to evaluate the performance of each models based on training and validation set. Two best-performing models were finally selected for ensemble modeling: model with conventional convolution (model 4) and the model with dilated convolution (model 5).

Table 1. Detailed parameters for all 5 3D U-Net models.

Model #	N	Loss type	Convolution
1	64, 64, 64	Cross-entropy	Conventional
2	64, 64, 64	Cross-entropy + Dice	Conventional
3	64, 96, 96	Cross-entropy	Conventional
4	64, 96, 96	Cross-entropy + Dice	Conventional
5	64, 96, 96	Cross-entropy + Dice	Dilated

The training time, including training two final models for deployment, was about 11 h per model and the training was performed on a Nvidia GeForce GTX 1080 Ti GPU with 11 Gb memory. TensorFlow framework was used to create and train the models. 600 epochs were used for training each model. Adam optimizer was used with a constant learning rate of 0.0005. Similarly, batch size of 1 was used and 3 layers were used. The number of input features used at the first layer was 48. During training of each epoch, one random patch was extracted for each subject where the orders of subject were randomly permuted. Finally, instance normalization was performed during testing, where each feature map was normalized using its mean and standard deviation.

2.4 Prediction Using Each Model

As discussed in the earlier section, using a large input size or whole images poses challenge of memory issue during training and so the patch-based training approach was implemented. The same challenge was faced during deployment, where the whole image cannot fit into the GPU memory. To combat the problem, sliding window approach was used to get the output for each testing subject. The sliding window approach, however, creates an issue of unstable predictions when sliding the window across the whole image without overlaps. When overlapping compared to the one without overlapping, the prediction time increases 8 times since it has to overlap in each dimension of the voxel. In deployment, stride size of 4 was used and the output probability was averaged. For each window, the original image and left-right flipped image were both predicted, and the average probability after flipping back the output of the flipped input was used as the output. Since ensemble modeling was performed after prediction using each model, raw probability output was saved for each model to the disk instead of the final labels which is produced after thresholding the probability output.

2.5 Ensemble Modeling

Ensemble method tend to have better predictive performance compared to individual models. Thus, the ensemble modeling process was performed to produce the final labels or binary mask based on the prediction of two different models. The raw probability maps of the two selected models were averaged. The final predictions were obtained by classifying voxels with an averaged probability above 0.5 as tumor. Since two classes are present, 0 for non-tumor and 1 for tumor, the final segmentation created the binary mask for tumor segmentation.

2.6   Image Post-processing

The last step before final submission was image post-processing. During post-processing, each final segmentation label (binary mask) was resampled using correct pixel spacing and origin based on the original CT 3D images after being cropped using the bounding box coordinates. Pixel spacing and origin were the main components used for resampling, since images were already cropped using bounding boxes during image pre-processing step. The post-processing was performed because of the requirement for the submission in the Challenge, where the output needs to be in the original CT resolution and cropped using respective bounding boxes.

3   Results

The whole training dataset of 201 cases or subjects were used in the training of final models. During preliminary training, the training dataset was split into 121 training subjects and 80 validation subjects. The training data comprised of medical images (CT and FDG-PET scans) from four different centers. For model selection, hyper parameter tuning was performed, and the performance of different models were evaluated using both dice score and loss score. Based on 5 models described in Table 1, Table 2 shows the results (training dice score, training loss and validation dice score) that were used during model selection process.

The two best performing models were selected and trained on the whole training dataset and used for final ensemble modeling

Table 2. Results for all 5 3D U-Net models.

Model #	Training dice	Training loss	validation dice
1	0.7226	0.011	0.5805
2	0.7136	0.360	0.4588
3	0.8578	0.006	0.6647
4	0.8885	0.130	0.7229
5	0.8880	0.130	0.7050

Based on their model performance, model 4 and model 5 were selected for ensemble modeling and the dice score was 0.7268 for validation split after ensemble. Finally, segmentation was performed on the testing data of 53 subjects which was from a different center as compared to the training set. The average Dice Similarity Coefficient (DSC), precision and recall for the test set were calculated by the MICCAI 2020 (HECKTOR) Challenge organizers after submission. The DSC, precision and recall were 0.6911, 0.7525 and 0.6928 respectively.

4   Discussion and Conclusions

In this paper we developed an automatic head and neck tumor segmentation method using an ensemble of 3D U-Nets. While the segmentation algorithm was developed for head and neck tumor, there is a potential to transfer the knowledge to other types of tumors/cancers and other similar studies. Information from FDG-PET and CT modalities were fused during pre-processing for accurate segmentation. During model selection, 5 models were trained with different patch size, loss type and convolution type as hyperparameters. The preliminary results showed improvement with larger patch size, where validation dice score was 0.4588 for patch size $64 \times 64 \times 64$ and 0.7229 dice score for patch size $64 \times 96 \times 96$. It also showed improvement when dice loss was added to cross-entropy loss, where validation dice score increased from 0.6647 to 0.7229. Similarly, during cross-validation, ensemble modeling performed better compared to the performance of two models separately. Furthermore, during testing, overlap in the sliding window with larger number of strides improved the performance during cross-validation. Thus, during deployment, stride of 4 was used for prediction in the testing set.

During model selection process, only 5 models were trained with hyperparameter tuning due to limitation in computation time. With further hyperparameter tuning, the performance could be improved. The two best performing models were selected and trained on the whole training dataset and used for final ensemble modeling. Experimenting with late fusion of two modalities (FDG-PET and CT) could have also shown if there would be a significant effect on performance and/or training time as compared to early fusion. Similarly, ensemble modeling with a greater number of models could also further improve the performance. In conclusion, we developed a patch-based 3D UNet for head and neck tumor segmentation with an ensemble of conventional and dilated convolution with early fusion of two modalities (FDG-PET and CT).

Acknowledgements. This project has been funded in whole or in part with Federal funds from the National Cancer Institute, National Institutes of Health, Department of Health and Human Services, under Contract No. 75N91020C00048.

References

1. Andrearczyk, V., et al.: Automatic segmentation of head and neck tumors and nodal metastases in PET-CT scans. In: Medical Imaging with Deep Learning, MIDL (2020)
2. Andrearczyk, V., et al.: Overview of the HECKTOR challenge at MICCAI 2020: automatic head and neck tumor segmentation in PET/CT. In: Andrearczyk, V., et al. (eds.) HECKTOR 2020. LNCS, vol. 12603, pp. 1–21. Springer, Cham (2021)
3. Ronneberger, O., Fischer, P., Brox, T.: U-Net: convolutional networks for biomedical image segmentation. In: Navab, N., Hornegger, J., Wells, W.M., Frangi, A.F. (eds.) MICCAI 2015. LNCS, vol. 9351, pp. 234–241. Springer, Cham (2015). https://doi.org/10.1007/978-3-319-24574-4_28

Tumor Segmentation in Patients with Head and Neck Cancers Using Deep Learning Based-on Multi-modality PET/CT Images

Mohamed A. Naser[(⊠)] ⓘ, Lisanne V. van Dijk ⓘ, Renjie He ⓘ,
Kareem A. Wahid ⓘ, and Clifton D. Fuller ⓘ

Department of Radiation Oncology, The University of Texas MD Anderson
Cancer, Houston, TX 77030, USA
manaser@mdanderson.org

Abstract. Segmentation of head and neck cancer (HNC) primary tumors on medical images is an essential, yet labor-intensive, aspect of radiotherapy. PET/CT imaging offers a unique ability to capture metabolic and anatomic information, which is invaluable for tumor detection and border definition. An automatic segmentation tool that could leverage the dual streams of information from PET and CT imaging simultaneously, could substantially propel HNC radiotherapy workflows forward. Herein, we leverage a multi-institutional PET/CT dataset of 201 HNC patients, as part of the MICCAI segmentation challenge, to develop novel deep learning architectures for primary tumor auto-segmentation for HNC patients. We preprocess PET/CT images by normalizing intensities and applying data augmentation to mitigate overfitting. Both 2D and 3D convolutional neural networks based on the U-net architecture, which were optimized with a model loss function based on a combination of dice similarity coefficient (DSC) and binary cross entropy, were implemented. The median and mean DSC values comparing the predicted tumor segmentation with the ground truth achieved by the models through 5-fold cross validation are 0.79 and 0.69 for the 3D model, respectively, and 0.79 and 0.67 for the 2D model, respectively. These promising results show potential to provide an automatic, accurate, and efficient approach for primary tumor auto-segmentation to improve the clinical practice of HNC treatment.

Keywords: PET · CT · Tumor segmentation · Head and neck cancer · Deep learning · Auto-contouring

1 Introduction

Head and neck cancer (HNC) affects over 50,000 individuals and has a mortality rate of over 10,000 annually [1]. A vast majority of HNC patients receive radiotherapy, which targets the tumor tissue with focused radiation beams from different directions, while trying to spare the surrounding tissues as much as possible [2]. Performed by the radiation oncologist, definite primary and lymph node tumor delineation dictates subsequent radiation dose optimization. The high prescribed dose is delivered to the segmented tumor, while limiting the dose directly surrounding the segmentation.

© Springer Nature Switzerland AG 2021
V. Andrearczyk et al. (Eds.): HECKTOR 2020, LNCS 12603, pp. 85–98, 2021.
https://doi.org/10.1007/978-3-030-67194-5_10

Inadequate tumor definition can therefore directly lead to under-dosage of the tumor, increasing treatment failure risk, or, in contrast, administering too much dose to the surrounding normal tissues. Adequate manual tumor segmentation is labor-intensive and subject to inter-observer variation [3–8]. Since, at present, CT tissue density information is needed for dose calculation, contours are defined on the CT, and often secondarily by ^{18}F-FDG Positron Emission Tomography (PET), providing additional information on the tissue's metabolic activity. Automatic segmentation of the primary tumor effectively utilizing the synergistic information from the PET and CT together is an unmet need to decrease the work-load of tumor delineation, as well as to decrease inter-variability between observers.

Deep learning (DL), an artificial intelligence subtype, is a strong tool for segmentation problems [9, 10]. DL techniques for segmentation applications on medical images for HNC radiotherapy purposes is a relative novel, yet emerging field [11]. An array of studies have peered into the difficult task of primary tumor segmentation with DL in single modality images, predominantly CT [12]. DL studies utilizing dual modalities, such as PET/CT [13–23], demonstrate the potential to outperform DL networks based on single image modalities [13, 21, 22, 24]. Likely due to the complex regional head and neck anatomy, PET/CT DL for HNC auto-contouring showed variable success, with dice similarity coefficients (DSC) ranging from 0.61 to 0.785 [13, 20–22]. These studies are often limited by small numbers of patients in the training and test datasets. The DL architectures for these studies vary, with 2D image (i.e. predictions made on a slice by slice basis) or 3D image approaches (i.e. predictions made by inputting the entire image volume) predominating.

The aim of this study was to develop and validate primary tumor auto-contouring with 2D/3D DL approaches that utilize PET and CT images simultaneously based on multi-institutional HNC data, as part of the MICCAI 2020: HECKTOR challenge.

2   Methods

We developed a deep learning model (Sect. 2.3) for auto-segmentation of primary tumors of HNC patients using co-registered ^{18}F-FDG PET and CT imaging data (Sect. 2.1). The ground truth manual segmentation of the tumors and the normalized imaging data (Sect. 2.2) were used to train the model (Sect. 2.4). The performance of the trained model for auto-segmentation was validated using a 5-fold cross validation approach (Sect. 2.5).

2.1   Imaging Data

The data set used in this paper, which was released by AIcrowd [25] for the HECK-TOR challenge at MICCAI 2020 [26], consists of co-registered ^{18}F-FDG PET and CT scans for 201 HNC patients, of which the majority were oropharyngeal cancer patients. All imaging data was paired with manual segmentations of the HN primary tumors, i.e. primary gross tumor volume (GTVp), which were considered as the ground truth, in Neuroimaging Informatics Technology Initiative (NIfTI) format.

2.2   Image Processing

To mitigate the variable resolution and size of the PET and CT image per patient, all images (i.e., PET, CT, and GTVp masks) were cropped to fixed bounding box volumes of size $144 \times 144 \times 144$ mm^3 in the x, y and z dimensions. These bounding boxes were provided with the imaging data (Sect. 2.1) by [25]. Then, the cropped images were resembled to a fixed image size of $144 \times 144 \times 96$ voxels. These specified number of voxels were chosen to match the maximum number of voxels found in the cropped CT images in the x, y, and z dimensions in all patients. The CT intensities were truncated in the range of $[-200, 200]$ Hounsfield Units (HU) to increase soft tissue contrast. The intensities of the truncated CT images were then rescaled to a $[-1, 1]$ range. The intensities of PET images were truncated between the 10th and 99th percentile to improve the images' contrast, and subsequently with z-normalization ([*intensity*-mean]/standard_deviation), resulting in a mean of zero and standard deviation of one for the entire cohort.

2.3   Segmentation Model Architecture

We developed 2D and 3D fully convolutional neural network (CNN) models based on the U-net architecture [27] and our previous 2D U-net model [28], using 4 convolution blocks in the encoding and decoding branches of the U-net. For each block, we used one convolution layer. The down sampling in the encoding branch was performed using a stride 2 convolution instead of max pooling layers to improve the model expressive ability through learning pooling operations compared to fixed pooling operations [29]. The up-sampling in the decoding branch was performed using convolution transpose layers which have been shown to be effective in previous studies [30–33]. Each convolution layer was directly followed by a batch normalization and a Leaky Relu activation layer; a Leaky Relu was chosen instead of Relu to mitigate the effect of improper model weight initialization and data normalization on the model training performance due to the "dying Relu" problem [34, 35]. The encoding and decoding blocks were linked using concatenation layers. Finally, the last layer was a Sigmoid activation layer. Figure 1 shows an illustration of the 3D U-net architecture proposed in this work. A similar architecture, but substituting the 3D with 2D convolution layers, was used to build the 2D U-net model. The batch normalization and Leaky Relu activation layers after each convolution layers were omitted from Fig. 1 for clarity. The number of filters used for the 4 convolution blocks were 16, 32, 48, 64, and 80 (Fig. 1). We maximized the number of filters (16 filters) in the first convolution block such that the data could be fit in GPU memory used for the model training, while an increment of 16 filters were used for the other convolution blocks. The total number of trainable parameters were 1,351,537 for the 3D model, and 452,113 for the 2D model.

2.4   Model Implementation

The processed PET and CT images (Sect. 2.2) were used as two inputs channels to the segmentation model (Sect. 2.3), resulting in an input layer size of [96, 144, 144, 2] for the 3D model and [144, 144, 2] for the 2D model which represent [z, y, x, channels] and [y, x, channels], respectively. The processed manual segmentation GTVp masks were used as the ground truth target to train the segmentation model. The processed images and masks were split into a training, validation, and test dataset (Sect. 2.5) and then used to train, validate, and test the segmentation model, accordingly. The optimizer used was 'Adam' with a learning rate of $5 * 10^{-5}$. The batch size was 1 for the 3D model and 96 for the 2D model. To minimize risk of over-fitting, data augmentation of the processed linked PET, CT, and mask images was implemented using a rotation range of $5°$, image scaling (i.e. zoom), intensity range shifting of 5%, and horizontal-flipping of images. The same random transformations were applied to the whole PET/CT/masks images for the 3D model per patient, while for the 2D model, each single image has different random transformations. The model performance metrics were the dice similarity coefficient (DSC), the recall or sensitivity, and precision or positive predictive value [28]. We note there is a class imbalance of tumor representation compared to normal tissue (i.e., the number of images that contain GTVp's is less than the number of images without GTVp's – i.e. normal tissue). This problem can lead to a low sensitivity in tumor identification by the model and lower the model performance for tumor segmentation. Therefore, to reduce the class imbalance effect, the model was trained using a loss function given as the summation of the loss function of DSC and a weighted Binary Cross Entropy (BCE) loss function as shown in Eqs. (1), (2), and (3).

$$\mathcal{L} = \mathcal{L}_{DSC} + \mathcal{L}_{BCE}, \tag{1}$$

$$\mathcal{L}_{DSC} = 1 - 2 \times \frac{\sum_i \mathbf{M}_i^{GT} \mathbf{M}_i^{Pred}}{\sum_i \mathbf{M}_i^{GT} + \sum_i \mathbf{M}_i^{Pred}}, \tag{2}$$

$$\mathcal{L}_{BCE} = \sum_i \mathbf{W}_i \left(\mathbf{M}_i^{GT} \log \mathbf{M}_i^{Pred} + \left(1 - \mathbf{M}_i^{GT} \right) \log \left(1 - \mathbf{M}_i^{Pred} \right) \right), \tag{3}$$

where $\mathbf{M}^{GT}$ and $\mathbf{M}^{Pred}$ are the ground truth and predicted tumor masks, respectively, and $\mathbf{W}$ is the sample-weight used to scale the loss for each image. The sample-weight is a function of the number of pixels in the provided ground truth manual segmentation mask as show in previous work [28]. The weight of the loss function that corresponds to tumor with larger cross-sectional area will be larger than that with smaller areas as well as normal tissue image. Figure 2 show an example of the sample-weight used to scale the loss of each image based on the size of the tumor of the image. The use of the weight-loss biases the model to focus on reducing the loss function more on images with larger tumor size compared to those with lower tumor size and normal tissue images and therefore improves the model sensitivity and overall model performance for tumor segmentation. The sample-weight is provided to the model as a second model input and it has the same size as the target ground truth tumor masks as shown in Fig. 1.

2.5 Model Training, Optimization and Validation

There is no available separate data that can be used to evaluate and validate the performance of the segmentation model. Therefore, we used a 5-fold (80% training and 20% validation) cross-validation approach where the 201 patients' imaging data and the corresponding ground truth tumor masks (Sect. 2.1) were split into 5 sets (Set 1 to Set 5). Each set contains imaging data of 40 patients randomly selected from the 201 patient dataset. The random split did not take into consideration the institutional sources of the data since the number of contributing patients varies significantly between these institutional centers (201 patients distributed as 72, 55, 18, and 56 from 4 different institutions); i.e. 4-fold cross validation based on patients from different institutions would provide significantly un-balanced train-validations sets and could lead to inaccurate estimation of the model performance. For each iteration of cross validation, each set of 40 patients serves as test data for the segmentation model trained using imaging data from the remaining 4 sets (i.e., 161 patients). Using this approach, the segmentation model was trained and tested 5 times. To estimate the number of epochs that should be used for training, the model was trained using 161 patients for training and 40 patients for validation, randomly selected from the 201 patients. The calculated loss using the validation data was used to obtain the maximum number of epochs before the model starts to overfit. In other words, when there is no further improvement of the loss evaluated in the optimization data. Using this approach, 50 epochs was estimated to be used for the 2D and 3D models training. Then, the segmentation model was trained for 50 epochs using 161 patients' data – 80% and tested using 40 patients' data – 20% 5 times. The overall DSC, recall, and precision values were obtained using the average of the mean DSC, recall, and precision values generated for the individual test data sets using the corresponding trained segmentation models. Subsequently, the model was trained one additional time, using 50 epochs, on the entire dataset (i.e., 201 patients' data) to generate the final model for the use of predicting the tumor masks the MICCAI challenge test set, i.e. a representation of other unseen datasets.

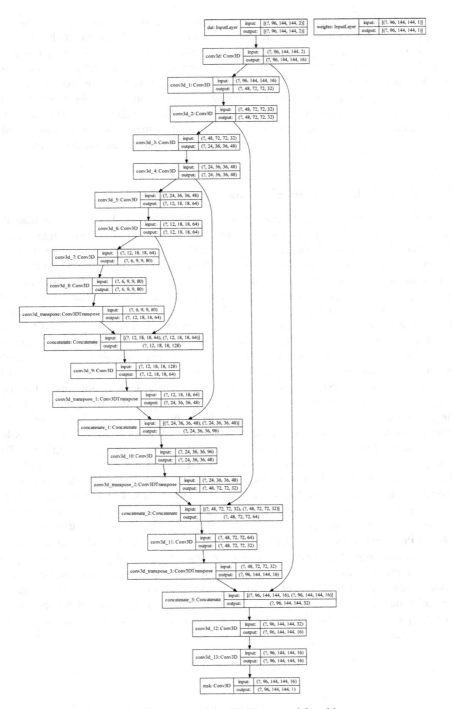

Fig. 1. An illustration of the 3D U-net model architecture.

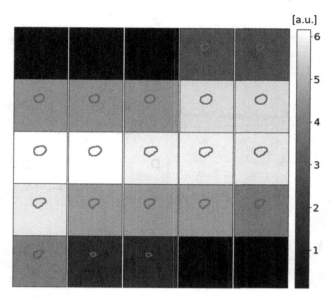

Fig. 2. An illustration of the sample-weight used to scale the BCE loss function for each image per patient based on the cross-sectional area of the tumor. The small squares show overlays of the tumor ground truth contours (red) and the cross-sectional images. Scale of the background grayscale color is the BCE weights. (Color figure online)

3   Results

The training performance of the model is illustrated in Fig. 3. The validation loss and DCS values do not show further improvements after epoch 45 for the 3D model and epoch 50 for the 2D model, consequently further model training led to model overfitting.

The DSC values' distributions obtained by the 3D and 2D segmentation models for the 5 test data sets using 50 epochs are illustrated in Fig. 4. The DSC median and mean values for the 3D model for Set 1 to Set 5 are 0.79, 0.79, 0.80, 0.78, and 0.79, respectively, and 0.72, 0.71, 0.70, 0.68, and 0.63, respectively. The DSC median and mean values for the 2D model for Set 1 to Set 5 are 0.81, 0.79, 0.80, 0.77, and 0.80, respectively, and 0.71, 0.70, 0.68, 0.63, and 0.61, respectively. The overall average (mean) values for the DSC, recall, and precision using all test data sets are presented in Table 1.

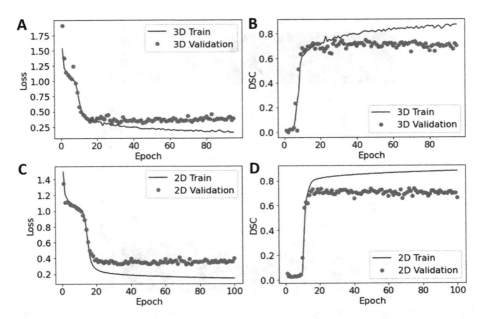

Fig. 3. The loss and DSC values as a function of epochs obtained during the 3D (A) and (B) and the 2D (C) and (D) model training.

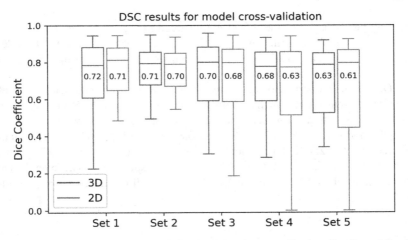

Fig. 4. Boxplots of the DSC distribution for the 5 test data sets (Set 1 to Set 5) used for the 3D and 2D segmentation model cross validation. The DSC mean values are given in the boxes and the lines inside the box refer to the DSC median values.

Table 1. 3D and 2D model performance metrics.

Model	DSC	Recall	Precision
3D	0.69 ± 0.03	0.75 ± 0.07	0.72 ± 0.03
2D	0.67 ± 0.04	0.71 ± 0.06	0.71 ± 0.03

To illustrate the performance of the segmentation model, samples of overlays of CT and PET images with the outlines of tumor masks using ground truth (red) and model segmentations (green) from the test data sets are shown in Fig. 5. The figure shows representative segmentation results for DSC values 0.51, 0.63, 0.80, 0.89, 0.92, and 0.94 which are below, comparable, and above the segmentation model's median DSC value of 0.79.

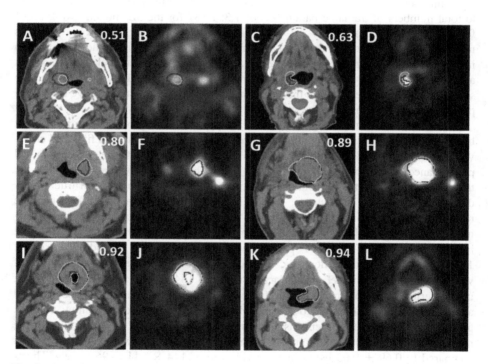

Fig. 5. 2D axial examples of overlays of the ground truth segmentations (red) and predicted segmentations (green) and CT images (first and third columns) and PET images (second and forth columns) with different 3D volumetric DSC values given at the right top. (Color figure online)

4   Discussion

As shown in Fig. 4 and Table 1, the 3D model outperforms the 2D model for all performance metrics. Specifically, the 3D model performance was superior to that of the 2D model for the mean DCS values for each test set (Set 1 to Set 5) (Fig. 4). Moreover, the mean DCS, recall, and precision values for all test sets (Table 1) are higher for the 3D model compared to the 2D model. As shown in Table 1, the mean DSC value of the 3D model (0.69) is larger than that of the 2D model (0.67). We performed a paired t-test on the 200 DSC values obtained from the 3D and 2D models (combining the DSC values of the 5 validation sets), resulting in a significant p-value of 0.043.

The image size of 144×144 used in the model training is relatively small compared to 512×512 and 256×256 usually used in the training of the standard U-net. Therefore, in the current 2D and 3D U-net models, to overcome overfitting, we used a small number of filters and 1 convolution layer for each convolution block which leads to total numbers of trainable parameters of 1,351,537 for the 3D model, and 452,113 for the 2D model compared to 7,759,521 for the standard U-net used in our previous model [28]. In addition, we used data augmentation to further mitigate the overfitting to improve the model performance. However, as seen in Fig. 3, the model starts to overfit after epoch 30 for the 3D model when trained on 161 patients. This indicates that the size of the data set used to train the model needs to be increased to improve the model performance and to mitigate the overfitting problem.

There are some limitations in the current approach. The proposed model has not been evaluated using an independent data set, instead, we performed a five-fold cross-validation approach (80% training and 20% validation) to estimate the expected model performance when applied to unseen test data. Therefore, it was assumed that the training data used by the model have a similar statistical representation of true unseen data which may not be accurate, especially if the unseen data has different image resolutions than the ones used to train the model. The model has been trained using images with $144 \times 144 \times 96$ voxels which are the maximum number of voxels found in the CT scans of the 201 patients within the provided bounding boxes of $144 \times 144 \times 144$ mm^3. Therefore, all images were up-sampled to that number of voxels. Training the model with a larger size (i.e., $144 \times 144 \times 144$) gives a lower model performance and increases overfitting as increasing the image size does not add additional useful information to the model describing the tumor. To show this, we trained the 3D model for 50 epochs using image size of $144 \times 144 \times 144$ voxels; the mean DSC values obtained from the 5 validation sets was lowered to 0.66 ± 0.04 compared to 0.69 ± 0.03 using images with size of $144 \times 144 \times 96$ voxels. Therefore, the model performance may be degraded when used to predict tumor masks using images with voxel size larger than $144 \times 144 \times 96$ as the input images will need to be down-sampled to a smaller size $144 \times 144 \times 96$ for masks' prediction.

The proposed models combine several novel features to improve performance such as a reduced U-net size, data augmentation, and a novel loss function for improving model sensitivity. For the 3D model, these features aid in achieving overall average median and mean DSC values of 0.79 and 0.69 respectively, comparable to a mean

DSC between radiation oncologists for HNC GTV delineation using PET/CT images (0.69) [36]. These features can be implemented in several other network architectures proposed for tumor segmentations such as ResNet [37], Inception [38], and DenseNet [39], which are worth investigating for future model improvement.

For the test data of the HECKTOR challenge, our 3D model was only able to achieve a DSC of 0.637. Interestingly, while DSC performance was generally lacking when compared to the other state of the art methods in the competition, our method was among the top models in precision (0.755). However, this came at a cost of low sensitivity (0.628). This indicates that the image voxels our model classified as tumor were very likely to be tumor, however, it subsequently was unable to detect many tumor voxels. While we are currently blinded to the ground truth contours of the test data, we can make some educated guesses on why our model did not generalize well in the test dataset. The test data includes several patients with images of higher resolutions than the ones used to train the model (i.e., the cropped images within the $144 \times 144 \times 144$ mm bounding boxes have sizes of $144 \times 144 \times 144$ voxels). These images were down-sampled to the size of $144 \times 144 \times 96$ voxels which is the size of the images used to train the model. Therefore, we expect a degradation of the model performance when used to predict the tumor masks of these images. The second reason for the discrepancy between the estimated DSC values using the training the test data could be due to the inaccurate estimation of the model performance on un-seen data using the proposed 5-fold cross validation the training data. Using 10-fold or larger internal validation strategies might provide a better estimation of the model performance on the test data.

5 Conclusion

This study presented a deep learning CNN model based on the U-net architecture to automatically segment primary tumors in HNC patients using co-registered FDG-PET/CT images. A combination of data normalization, dual input channel integration of PET and CT data, data augmentation, and the use of a loss function that combines contributions from the DSC while weighting BCE resulting in a promising performance of 3D tumor auto-segmentation with overall average median and mean cross-validation DSC values of 0.79 and 0.69, respectively. While our 3D model showed lower performance on a held-out test dataset, our methods are still useful for the auto-contouring community to incorporate and improve upon.

Acknowledgements. M.A.N. is supported by a National Institutes of Health (NIH) Grant (R01 DE028290-01). K.A.W. is supported by a training fellowship from The University of Texas Health Science Center at Houston Center for Clinical and Translational Sciences TL1 Program (TL1 TR003169). C.D.F. received funding from the National Institute for Dental and Cranio-facial Research Award (1R01DE025248-01/R56DE025248) and Academic-Industrial Partnership Award (R01 DE028290), the National Science Foundation (NSF), Division of Mathematical Sciences, Joint NIH/NSF Initiative on Quantitative Approaches to Biomedical Big Data (QuBBD) Grant (NSF 1557679), the NIH Big Data to Knowledge (BD2K) Program of the National Cancer Institute (NCI) Early Stage Development of Technologies in Biomedical Computing, Informatics, and Big Data Science Award (1R01CA214825), the NCI Early Phase

Clinical Trials in Imaging and Image-Guided Interventions Program (1R01CA218148), the NIH/NCI Cancer Center Support Grant (CCSG) Pilot Research Program Award from the UT MD Anderson CCSG Radiation Oncology and Cancer Imaging Program (P30CA016672), the NIH/NCI Head and Neck Specialized Programs of Research Excellence (SPORE) Developmental Research Program Award (P50 CA097007) and the National Institute of Biomedical Imaging and Bioengineering (NIBIB) Research Education Program (R25EB025787). He has received direct industry grant support, speaking honoraria and travel funding from Elekta AB.

References

1. Siegel, R.L., Miller, K.D., Jemal, A.: Cancer statistics, 2020. CA Cancer J. Clin. **70**, 7–30 (2020). https://doi.org/10.3322/caac.21590
2. Rosenthal, D.I., et al.: Beam path toxicities to non-target structures during intensity-modulated radiation therapy for head and neck cancer. Int. J. Radiat. Oncol. Biol. Phys. **72**, 747–755 (2008). https://doi.org/10.1016/j.ijrobp.2008.01.012
3. Vorwerk, H., et al.: Protection of quality and innovation in radiation oncology: the prospective multicenter trial the German Society of Radiation Oncology (DEGRO-QUIRO study). Strahlentherapie und Onkol. **190**, 433–443 (2014)
4. Riegel, A.C., et al.: Variability of gross tumor volume delineation in head-and-neck cancer using CT and PET/CT fusion. Int. J. Radiat. Oncol. Biol. Phys. **65**, 726–732 (2006). https://doi.org/10.1016/j.ijrobp.2006.01.014
5. Rasch, C., Steenbakkers, R., Van Herk, M.: Target definition in prostate, head, and neck. Semin. Radiat. Oncol. **15**, 136–145 (2005). https://doi.org/10.1016/j.semradonc.2005.01.005
6. Breen, S.L., et al.: Intraobserver and interobserver variability in GTV delineation on FDG-PET-CT images of head and neck cancers. Int. J. Radiat. Oncol. Biol. Phys. **68**, 763–770 (2007). https://doi.org/10.1016/j.ijrobp.2006.12.039
7. Segedin, B., Petric, P.: Uncertainties in target volume delineation in radiotherapy–are they relevant and what can we do about them? Radiol. Oncol. **50**, 254–262 (2016)
8. Anderson, C.M., et al.: Interobserver and intermodality variability in GTV delineation on simulation CT, FDG-PET, and MR images of head and neck cancer. Jacobs J. Radiat. Oncol. **1**, 6 (2014)
9. Guo, Y., Liu, Yu., Georgiou, T., Lew, M.S.: A review of semantic segmentation using deep neural networks. Int. J. Multimed. Inf. Retr. **7**(2), 87–93 (2017). https://doi.org/10.1007/s13735-017-0141-z
10. Garcia-Garcia, A., Orts-Escolano, S., Oprea, S., Villena-Martinez, V., Garcia-Rodriguez, J.: A Review on Deep Learning Techniques Applied to Semantic Segmentation (2017)
11. Boldrini, L., Bibault, J.-E., Masciocchi, C., Shen, Y., Bittner, M.-I.: Deep learning: a review for the radiation oncologist. Front. Oncol. **9**, 977 (2019)
12. Cardenas, C.E., Yang, J., Anderson, B.M., Court, L.E., Brock, K.B.: Advances in auto-segmentation. Semin. Radiat. Oncol. **29**, 185–197 (2019). https://doi.org/10.1016/j.semradonc.2019.02.001
13. Oreiller, V.A.V., Vallieres, M., Castelli, J., Boughdad, H.E.M.J.S., Adrien, J.O.P.: Automatic Segmentation of Head and Neck Tumors and Nodal Metastases in PET-CT scans (2020). http://proceedings.mlr.press/v121/andrearczyk20a.html
14. Li, L., Zhao, X., Lu, W., Tan, S.: Deep learning for variational multimodality tumor segmentation in PET/CT. Neurocomputing **392**, 277–295 (2020)
15. Leung, K.H., et al.: A physics-guided modular deep-learning based automated framework for tumor segmentation in PET images. arXiv Preprint arXiv:2002.07969 (2020)

16. Kawauchi, K., et al.: A convolutional neural network-based system to classify patients using FDG PET/CT examinations. BMC Cancer **20**, 1–10 (2020). https://doi.org/10.1186/s12885-020-6694-x

17. Zhong, Z., et al.: 3D fully convolutional networks for co-segmentation of tumors on PET-CT images. In: 2018 IEEE 15th International Symposium on Biomedical Imaging (ISBI 2018), pp. 228–231. IEEE (2018)

18. Jemaa, S., Fredrickson, J., Carano, R.A.D., Nielsen, T., de Crespigny, A., Bengtsson, T.: Tumor segmentation and feature extraction from whole-body FDG-PET/CT using cascaded 2D and 3D convolutional neural networks. J. Digit. Imaging. **33**, 888–894 (2020). https://doi.org/10.1007/s10278-020-00341-1

19. Zhao, X., Li, L., Lu, W., Tan, S.: Tumor co-segmentation in PET/CT using multi-modality fully convolutional neural network. Phys. Med. Biol. **64**, 15011 (2018)

20. Huang, B., et al.: Fully automated delineation of gross tumor volume for head and neck cancer on PET-CT using deep learning: a dual-center study. Contrast Media Mol. Imaging **2018**, 1–12 (2018). https://pubmed.ncbi.nlm.nih.gov/30473644/

21. Moe, Y.M., et al.: Deep learning for automatic tumour segmentation in PET/CT images of patients with head and neck cancers. arXiv Preprint arXiv:1908.00841 (2019)

22. Guo, Z., Li, X., Huang, H., Guo, N., Li, Q.: Deep learning-based image segmentation on multimodal medical imaging. IEEE Trans. Radiat. Plasma Med. Sci. **3**, 162–169 (2019)

23. Jin, D., et al.: Accurate esophageal gross tumor volume segmentation in PET/CT using two-stream chained 3D deep network fusion. In: Shen, D., Liu, T., et al. (eds.) MICCAI 2019. LNCS, vol. 11765, pp. 182–191. Springer, Cham (2019). https://doi.org/10.1007/978-3-030-32245-8_21

24. Zhou, T., Ruan, S., Canu, S.: A review: deep learning for medical image segmentation using multi-modality fusion. Array **3**, 100004 (2019)

25. AIcrowd MICCAI 2020: HECKTOR Challenges. https://www.aicrowd.com/challenges/miccai-2020-hecktor. Accessed 07 Sept 2020

26. Andrearczyk, V., et al.: Overview of the HECKTOR challenge at MICCAI 2020: automatic head and neck tumor segmentation in PET/CT. In: Andrearczyk, V., et al. (eds.) HECKTOR 2020. LNCS, vol. 12603, pp. 1–21. Springer, Cham (2021)

27. Ronneberger, O., Fischer, P., Brox, T.: U-Net: convolutional networks for biomedical image segmentation. In: Navab, N., Hornegger, J., Wells, W.M., Frangi, A.F. (eds.) MICCAI 2015. LNCS, vol. 9351, pp. 234–241. Springer, Cham (2015). https://doi.org/10.1007/978-3-319-24574-4_28

28. Naser, M.A., Deen, M.J.: Brain tumor segmentation and grading of lower-grade glioma using deep learning in MRI images. Comput. Biol. Med. **121**, 103758 (2020). https://doi.org/10.1016/j.compbiomed.2020.103758

29. Springenberg, J.T., Dosovitskiy, A., Brox, T., Riedmiller, M.: Striving for simplicity: the all convolutional net. arXiv Preprint arXiv:1412.6806 (2014)

30. Noh, H., Hong, S., Han, B.: Learning deconvolution network for semantic segmentation. In: Proceedings of the IEEE International Conference on Computer Vision, pp. 1520–1528 (2015). https://doi.org/10.1109/ICCV.2015.178

31. Long, J., Shelhamer, E., Darrell, T.: Fully convolutional networks for semantic segmentation. In: Proceedings of the IEEE Computer Society Conference on Computer Vision and Pattern Recognition, pp. 3431–3440 (2015). https://doi.org/10.1109/CVPR.2015.7298965

32. Badrinarayanan, V., Kendall, A., Cipolla, R.: SegNet: a deep convolutional encoder-decoder architecture for image segmentation. IEEE Trans. Pattern Anal. Mach. Intell. **39**, 2481–2495 (2017). https://doi.org/10.1109/TPAMI.2016.2644615

33. Krizhevsky, A., Sutskever, I., Hinton, G.E.: ImageNet classification with deep convolutional neural networks. Commun. ACM **60**, 84–90 (2017). https://doi.org/10.1145/3065386

34. Maas, A.L., Hannun, A.Y., Ng, A.Y.: Rectifier nonlinearities improve neural network acoustic models. In: Proceedings of ICML, p. 3 (2013)
35. Xu, B., Wang, N., Chen, T., Li, M.: Empirical evaluation of rectified activations in convolutional network. arXiv Preprint arXiv:1505.00853 (2015)
36. Gudi, S., et al.: Interobserver variability in the delineation of gross tumour volume and specified organs-at-risk during IMRT for head and neck cancers and the impact of FDG-PET/CT on such variability at the primary site. J. Med. Imaging Radiat. Sci. **48**, 184–192 (2017). https://doi.org/10.1016/j.jmir.2016.11.003
37. Zhang, Q., Cui, Z., Niu, X., Geng, S., Qiao, Y.: Image segmentation with pyramid dilated convolution based on ResNet and U-Net. In: Liu, D., Xie, S., Li, Y., Zhao, D., El-Alfy, E.S. (eds.) Lecture Notes in Computer Science (including subseries Lecture Notes in Artificial Intelligence and Lecture Notes in Bioinformatics), pp. 364–372. Springer, Cham (2017). https://doi.org/10.1007/978-3-319-70096-0_38
38. Szegedy, C., Vanhoucke, V., Ioffe, S., Shlens, J., Wojna, Z.: Rethinking the inception architecture for computer vision. In: Proceedings of the IEEE Computer Society Conference on Computer Vision and Pattern Recognition, pp. 2818–2826. IEEE Computer Society (2016). https://doi.org/10.1109/CVPR.2016.308
39. Jégou, S., Drozdzal, M., Vazquez, D., Romero, A., Bengio, Y.: The One Hundred Layers Tiramisu: Fully Convolutional DenseNets for Semantic Segmentation (2017). https://openaccess.thecvf.com/content_cvpr_2017_workshops/w13/html/Jegou_The_One_Hundred_CVPR_2017_paper.html

GAN-Based Bi-Modal Segmentation Using Mumford-Shah Loss: Application to Head and Neck Tumors in PET-CT Images

Fereshteh Yousefirizi[1] and Arman Rahmim[1,2(✉)]

[1] Department of Integrative Oncology, BC Cancer Research Institute,
Vancouver, BC, Canada
arahmim@bccrc.ca

[2] Departments of Radiology and Physics, University of British Columbia,
Vancouver, BC, Canada

Abstract. A deep model based on SegAN, a generative adversarial network (GAN) for medical image segmentation, is proposed for PET-CT image segmentation, utilizing the Mumford-Shah (MS) loss functional. An improved V-net is used for the generator network, while the discriminator network has a similar structure to the encoder part of the generator network. The improved polyphase V-net style network can help preserve boundary details unlike conventional V-net. A multi-term loss function consisting of MS loss and multi-scale mean absolute error (MAE) was designed for the training scheme. Using the complementary information extracted via MS loss helps improve supervised segmentation task by regularizing pixel/voxel similarities. MAE as the semantic term of loss function compensates for probable subdivisions into intra-tumor regions. The proposed method was applied for automatic segmentation of head and neck tumors and nodal metastases based on the bi-modal information from PET and CT images, which can be valuable for automated metabolic tumor volume measurements as well as radiomics analyses. The proposed bi-modal method was trained on 201 PET-CT images from four centers and was tested on 53 cases from a different center. The performance of our proposed method, independently evaluated in the HECKTOR challenge, achieved average Dice score coefficient (DSC) of 67%, precision of 73% and recall of 72%.

Keywords: Head and neck cancer · PET-CT · GAN · 3D segmentation · Mumford-Shah loss

1 Introduction

PET-CT based image segmentation is an essential step for radiomics analyses used for improved quantitative assessment of cancer [1], including head and neck cancers [2]. As an example, such segmentation can enable quantification of metabolic tumor volume (MTV) [3] that is known to have significant predictive and prognostic value beyond more easy-to-measure metrics (e.g. maximum standard uptake value; SUVmax).

© Springer Nature Switzerland AG 2021
V. Andrearczyk et al. (Eds.): HECKTOR 2020, LNCS 12603, pp. 99–108, 2021.
https://doi.org/10.1007/978-3-030-67194-5_11

Manual delineation of tumor regions is time-consuming and prone to errors [4] with considerable intra- [5] and inter-observer variability [6, 7]. Given that tumor contours are not always the same in PET vs. CT images [8], multi-modality segmentation methods have been implemented to utilize the functional and metabolic information of PET images and anatomical localization of CT images simultaneously [7, 9–12] or separately [13, 14]. The latter methods have inherent limitations, and instead, bi-modal simultaneous segmentation can enable more meaningful incorporation of functional and anatomical data [9]. Some existing multi-modality PET-CT segmentation techniques are time-consuming or require pre- and/or post-processing steps [15] that are acceptable to some extent. In order to use bi-modal information of PET and CT images, some proposed methods used multiple-channel inputs (i.e. via early fusion [4]) to the convolutional network. In other words, PET and CT images are implicitly fused as proposed by Bradshaw et al. [16] and Zhang et al. [17]. Using separate branches, specific to each modality, was also proposed for bi-modal segmentation [8,13]. Spatially-varying fusion of PET-CT images was recently proposed by Kumar et al. [9] to quantify the explicit fusion weights of PET and CT images considering the priority of each modality in different regions.

Automated segmentation of head and neck tumors has gained attention recently. Moe et al. [18] used a U-net model, and PET and CT windowing, for tumor and metastatic lymph node segmentation. Jin et al. [7] proposed early and late fusion pipelines based on a 3D deep network for esophageal gross tumor volume delineation in PET and (radiotherapy) CT images. Andrearczyk et al. [4] used 2D and 3D V-net [19] for PET only, CT only and PET-CT late fusion segmentation of oropharynx tumors. Their best results were obtained by 2D V-net on PET-CT late fusion data [4].

In the current study, we propose a 3D network for bi-modal PET/CT segmentation based on SegAN, a generative adversarial network (GAN) [20] for medical image segmentation [21] (Sect. 2.2). GAN has shown excellent performance for image segmentation with low amount of labeled data [22]. We additionally propose a multi-term loss function by adding Mumford-Shah (MS) loss approach of Kim and Ye [23] to the multi-scale loss function of SegAN, in order to improve the segmentation results, especially in the presence of weak ground truth, towards improved generalizability of our model (Sect. 2.3). Regions with similar pixel/voxel values are captured by MS loss that reduces the dependency on time-consuming and prone-to-error manual delineations. A 2D-equivalent architecture of our proposed method was also developed and applied to the data.

In the following sections, we first introduce the data provided by the MICCAI 2020 HEad and neCK TumOR (HECKTOR) segmentation challenge [24]. Our proposed methods and training scheme are additionally explained. The results are then presented, followed by discussion and conclusion.

2   Materials and Methods

2.1   Dataset

CT and PET images and the primary gross tumor volume (GTV) of 201 patients with oropharynx tumors were used for training, gathered from four centers; CHGJ (55 cases), CHMR (18 cases), CHUM (56 cases) and CHUS (72 cases) originating from study by Vallieres et al. [2]. CT and PET images of 53 patients used for independent testing were from a different center (CHUV). All the training and test data were in NIFTI format. Voxel dimensions of the CT and PET images were resampled to $1 \times 1 \times 1$ mm^3 using trilinear interpolation to produce isotropic voxel dimensions. The PET and CT images were cropped to $144 \times 144 \times 144$ ($144 \times 144 \times 1$ for 2D approach), containing tumors areas. The CT windowing previously reported to improve the results [4, 18] had no considerable effects on our results; consequently we did not use this windowing (clipping) on Hounsfield Unit values in our segmentation pipeline. Images from both modalities (PET and CT) were normalized in the range of [0, 1].

2.2   Proposed Method

PET and CT images are entered into the V-net style Segmentor network as multiple input channels. The Segmentor produces a probability label map as output. Multiplication of the original image and output produces the predicted mask, and multiplication of image and ground truth yields the ground truth mask. The predicted and ground truth masks are inputs to the Critic network.

The single scalar "real/fake" output of discriminator network in the classic version of GAN [20] was reported to produce unstable and insufficient gradient feedback to the network, which may be problematic in segmentation applications that needs dense and pixel (voxel) level labeling [21]. Xue et al. proposed an end-to-end adversarial network SegAN [21] applying a multi-scale L1 loss function between the predicted mask (pixel-wise multiplication of predicted label map and original image) and the target mask (pixel-wise multiplication of ground truth label map and original image). Luc et al. [25] proposed a weighted sum of two loss functions for an adversarial network that discriminates the generated segmentations and the ground truths. By contrast, we propose a multi-term loss function in the GAN-based model. We used a 3D V-net style architecture modified by polyphase decomposition proposed by Kim et al. [26] to preserve the high frequency information of boundaries as the generator network (Segmentor). It has been shown that this property improves the segmentation accuracy of U-net [26].

Based on the theory of deep convolutional framelets [27], it was shown by Kim et al. [26] that polyphase U-Net satisfies the frame condition that helps retain the high frequency detailed components. The pooling and unpooling layers of the V-Net, in our proposed method was modified based on this fact by using polyphase decomposition. The four neighbor voxels of the input images are decomposed into the 4 channel data with reduced size at the pooling layer. At the unpooling layer, the 4 reduced size channels are then composed an expanded channel [26].

A framework similar to the encoder part of the generator was used for the discriminator network (Critic) [21]. During training, the generator network is fixed for one step and the discriminator network is trained repeatedly inspired by SegAN [21]. The architecture of our proposed method is shown in Fig. 1. Our contribution has been to utilize the GAN model in a SegAN based framework using V-Net, with a new multiscale loss in combination with MS loss to capture regions with differing size, shape and heterogeneity. The proposed method is also capable of retaining accurate boundaries due to the polyphase architecture.

2.3   Training Scheme

Loss functions used in conventional GANs measure the distance between the distributions of the generated vs. real data, as GANs try to replicate a probability distribution [20]. Our proposed loss function has two terms; the MS loss, l_{MS}, and the adversarial model loss, l_A:

$$l(\theta_S, \theta_C) = l_{MS} + l_A \tag{1}$$

where θ_s and θ_C are the parameters of the Segmentor and Critic networks, respectively, to be learned in the min-max game. The first term is MS loss functional that is calculated based on the pixel similarity between output (y_n) and original image (x_n), and was introduced by Kim et al. for weakly-supervised or unsupervised semantic segmentation applications [23].

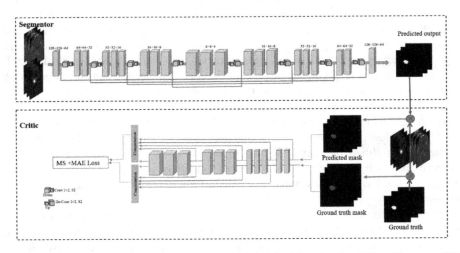

Fig. 1. The architecture of the proposed method inspired by the SegAN architecture [21]. A modified 3D V-net style architecture is used the generator network (Segmentor). The discriminator (Critic) network is similar to the encoder part of the generator. Predicted mask is calculated by voxel-wise multiplications of original image and predicted output. Ground truth mask is also calculated by pixel-wise multiplication of original image and ground truth. Predicted mask and ground truth masks are the two inputs to the Critic network.

The minimization of the MS functional is simply based on back-propagation, and there is no need for the computational burden of Euler-Lagrangian evolution; the key is the fact that the softmax layer of CNN is similar to the differentiable approximation of the characteristic function of MS functional for image segmentation. Consequently it can be minimized using the deep network without added computational burden [23].

In the case of supervised segmentation, the MS term is used as the regularized function to improve segmentation results, and is given by:

$$l_{MS}(\Theta; x) = \sum_{n=1}^{N} \int_{\Omega} |x(r) - c_n|^2 y_n(r)dr + \lambda \sum_{n=1}^{N} \int_{\Omega} |\nabla y_n(r)|dr \qquad (2)$$

where $x(r)$ and $y_n(r)$ are the input and output of the softmax layer, respectively, Θ refers to the learnable parameters, and c_n is the average pixel value defined as follows:

$$c_n(\Theta) = \frac{\int_{\Omega} x(r)y_n(r; \Theta)dr}{\int_{\Omega} y_n(r; \Theta)dr} \qquad (3)$$

Adversarial training is based on the second term of loss function, l_A (Eq. (1)). The adversarial term, adapted from SegAN [21], is defined as follows:

$$l_A = \frac{1}{N} \sum_{n=1}^{N} l_{MAE}(f_C(x_n \circ y_n), f_C(x_n \circ g_n)) \qquad (4)$$

where l_{MAE} is the mean absolute error, g_n is the ground truth, $(x_n \circ y_n)$ and $(x_n \circ g_n)$ denote voxel-wise multiplications of original image and output (predicted mask), and of original image and ground truth (ground truth mask), respectively. $f_C(x)$ shows the features that are hierarchically extracted in different layers (with different scales) of the Critic network (Fig. 1) [21]. Multi-scale spatial differences between the voxels of predicted masks and ground truth masks in terms of mean absolute error are calculated hierarchically. The adversarial loss is minimized with respect to the parameters of Segmentor network, θ_S while it is maximized with respect to the parameters of the Critic network, θ_C. As training process evolves, the Segmentor network learns to produce outputs that are close to ground truth.

Our network was trained using the Adam optimizer [28] with a learning rate of 0.00001 and 350 iterations, and a batch size of 4 for 3D implementation. For the 2D network, the number of iterations and batch size were 1000 and 12, respectively. Training was completed using an NVIDIA Tesla V100 GPUs 16 GB after 6 h for 3D and 10 h for 2D network. The inference times were 20 s for both 3D and 2D networks.

3   Results

The performance of our proposed method was independently evaluated on test data in the HECKTOR challenge, and achieved an average Dice score coefficient (DSC) of 67%, precision of 73% and recall of 72%. Besides, we evaluated our 3D model in a separate training scheme via a leave-one-center-out cross-validation, similar to [4] on the training data to compare its performance to 2D and 3D V-net (using NiftyNet [29]). Given variations due to initializations, this cross-validation effort was performed 10 times, with mean DSC values (and Precision) reported in Table 1. The 95% Confidence Intervals (CI) were also computed on all cases across the 10 runs regardless of the specific centers (the CIs were calculated as follows: Confidence interval = sample mean ± margin of error).

Table 1. Quantitative evaluation of proposed bi-modal 3D and 2D GAN + MS networks compared to 3D and 2D V-Nets. Average DSC and Precision (%) are reported with 95% CIs. (DSC: mean dice similarity)

Method	DSC	Precision
Proposed 3D GAN + MS	69% ± 2.1	78% ± 2.2
Proposed 2D GAN + MS	65% ± 2.2	74% ± 2.1
3D V-net (Dice + Cross-entropy)	62.3% ± 2.2	62.6% ± 2.4
2D V-net (Dice + Cross-entropy)	62% ± 2.1	61% ± 2.3

Our proposed 3D GAN + MS model resulted in relatively higher mean Dice similarity with a variability across different centers (69% ± 2.1) and higher precision (78% ± 2.2) when utilizing our dataset (the separate test data were are not considered in this comparison). Andrearczyk et al. reported [4] improved segmentation results of bi-modal 3D and 2D V-net using NiftyNet [29] compared to PET-only or CT-only based segmentations.

The number of batch size, iterations and the type of optimizer and learning rate for 2D and 3D V-net (NiftyNet) were selected based on [4] and the training process was repeated for 2D and 3D model; we do not claim that we used exactly the same model for comparison though we made best effort to replicate the network. It is worth mentioning that performances of 3D and 2D V-net in our implementation were higher than those reported on similar data in [4].

Figure 2 depicts images with our proposed 2D and 3D GAN + MS models vs. the ground truth to illustrate challenges about manual annotations. Although 3D model resulted in higher DSC value, but qualitatively we can visually inspect that 2D model may actually delineate the tumor region more accurately. Since DSC, precision and other segmentation measures are calculated based on the manual 'ground truth', variability in the latter can impact quality of the models; in fact having more robust 'ground truth' data can enable generation of robust models with less training data [30].

Figure 3 shows ability of the proposed 3D GAN + MS model to capture the tumor region in the presence of low quality labels. By contrast, 2D and 3D V-net could not segment any tumor region in this particular PET/CT image. MS loss, by contrast, was able to do so. This underscores the ability of MS loss to segment real tumor regions in the presence of noisy/unreliable labels and/or data, especially in the case of 3D model in our proposed method.

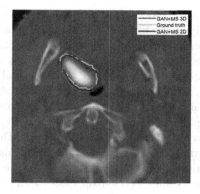

Fig. 2. The axial view of PET/CT image and the segmented region by 3D GAN + MS (red contour) (DSC = 83%) and 2D GAN + MS (blue contour) (DSC = 80%). Green contour depicts ground truth. (Color figure online)

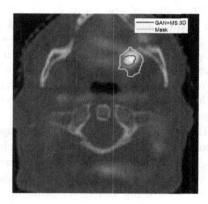

Fig. 3. Bi-modal segmentation result of 3D GAN + MS is overlaid on the axial PET/CT image (red contour (DSC = 28%)). 2D GAN + MS, 3D and 2D V-net did not extract this tumor. Green contour depicts the ground truth. (Color figure online)

4 Discussions and Conclusion

Comparing the segmentation results of our 2D and 3D GAN + MS methods, using a leave-one-center-out scheme, 3D model outperforms 2D model in the present head and neck PET/CT data (Table 1). Previous experiments on segmentation of data similar to our training data by Andrearczyk et al. [4] revealed that 2D V-net slightly outperformed the 3D V-net model. In spite of the 'curse of dimensionality' problem of 3D segmentation that is caused by data sparsity and probable overfitting, it has been shown that 3D segmentation methods based on deep models could overcome these limitations and even outperform 2D and 2.5D methods [31]. In the present study, better performance of the 3D model (Fig. 3) can be attributed to taking into consideration the 3D spatial information of the image data especially in the tumor regions in the 3D model compared to the 2D model, and more importantly using MS loss for better extraction of the tumor region and mitigate the limitation of the weak ground truths.

Additional investigation using different multi-modality datasets is needed to further evaluate model robustness. Learning based on ground truths made by multiple annotators may help the method to be more robust [4]. As manual annotation is a difficult task and prone to errors, with intra and interobserver variability and inconsistency across oncologists [4, 6], aiming to make automated segmentations "close to oncologist level" may not always capture the true tumor region and GTV. In other words, using a larger dataset may not always result in better performance. The model performance is always limited by the consistency of the ground truth [30]. This motivates development of multi-modal segmentation models with altered loss functions, such as MS, that are able to capture regions with similar pixel/voxel values with regularized contour length to reduce the dependency on variable manual annotations. When outcome data is available, then it is possible to investigate how deep learning based methods truly compare with manual segmentation, in predictive and prognostic tasks [32].

Our work has a number of limitations and potential for further developments. We utilized dice and dice + cross-entropy loss functions for adversarial training, which worked well, though the segmentation results of the trained network by MAE loss depicted higher scores (reported in Table 1 as 3D V-net (dice + cross-entropy) and 2D V-net (dice + cross-entropy). In the supervised training scheme, MS loss should be weighted lower than adversarial term, and the coefficient can be adaptively changed based on reliability of ground truth. We used a fixed coefficient ($\alpha = 0.001$) for all training data. In future efforts, we aim to improve our proposed model to attain further improved segmentation scores.

We used the bi-modal information of PET and CT images as multiple-channel inputs to the proposed convolutional network. Spatially-weighted fusion of PET and CT images based on the priority of each modality in different regions [9] will be considered in our future work to improve our multi-modality segmentation results.

There are limited studies on using GANs for semantic segmentation [25] and especially for medical image segmentation, an exception being SegAN [21] that was especially designed for this aim, which we further explored and expanded it in our present work. Recently Kim et al. [26] used cycle-consistent GAN (CycleGAN) [33] for liver tumor segmentation in CT images [26]. We are further exploring this network for bi-modal segmentation. Further studies are needed to investigate generalizability of our proposed models to segment head and neck tumors in other large multi-center datasets, as well as application to other types of tumors.

Acknowledgments. The authors gratefully acknowledge Dr. Ivan Klyuzhin for his valuable support and feedback. This research was supported in part through computational resources and services provided by Microsoft and the Vice President Research and Innovation at the University of British Columbia.

References

1. Lambin, P., et al.: Radiomics: extracting more information from medical images using advanced feature analysis. Eur. J. Cancer **48**(4), 441–446 (2012)
2. Vallieres, M., et al.: Radiomics strategies for risk assessment of tumour failure in head-and-neck cancer. Sci. Rep. **7**(1), 1–14 (2017)
3. Im, H.-J., et al.: Current methods to define metabolic tumor volume in positron emission tomography: which one is better? Nucl. Med. Mol. Imaging **52**(1), 5–15 (2017). https://doi.org/10.1007/s13139-017-0493-6
4. Andrearczyk, V., et al.: Automatic segmentation of head and neck tumors and nodal metastases in PET-CT scans. In: Medical Imaging with Deep Learning MIDL, Montreal (2020)
5. Starmans, M.P., et al.: Radiomics: data mining using quantitative medical image features. In: Handbook of Medical Image Computing and Computer Assisted Intervention, pp. 429–456. Elsevier (2020)
6. Gudi, S., et al.: Interobserver variability in the delineation of gross tumour volume and specified organs-at-risk during IMRT for head and neck cancers and the impact of FDG-PET/CT on such variability at the primary site. J. Med. Imaging Radiat. Sci. **48**(2), 184–192 (2017)
7. Jin, D., et al.: Accurate Esophageal Gross Tumor Volume segmentation in PET/CT using two-stream chained 3D deep network fusion. In: Shen, D., et al. (eds.) MICCAI 2019. LNCS, vol. 11765, pp. 182–191. Springer, Cham (2019). https://doi.org/10.1007/978-3-030-32245-8_21
8. Zhong, Z., et al.: Simultaneous cosegmentation of tumors in PET-CT images using deep fully convolutional networks. Med. Phys. **46**(2), 619–633 (2019)
9. Kumar, A., et al.: Co-learning feature fusion maps from PET-CT images of lung cancer. IEEE Trans. Med. Imaging **39**(1), 204–217 (2019)
10. Li, L., et al.: Deep learning for variational multimodality tumor segmentation in PET/CT. Neurocomputing **392**, 277–295 (2019)
11. Zhao, Y., et al.: Deep neural network for automatic characterization of lesions on 68 Ga-PSMA-11 PET/CT. Eur. J. Nucl. Med. Mol. Imaging **47**(3), 603–613 (2020)
12. Han, D., et al.: Globally optimal tumor segmentation in PET-CT images: a graph-based co-segmentation method. In: Székely, G., Hahn, H.K. (eds.) IPMI 2011. LNCS, vol. 6801, pp. 245–256. Springer, Heidelberg (2011). https://doi.org/10.1007/978-3-642-22092-0_21
13. Teramoto, A., et al.: Automated detection of pulmonary nodules in PET/CT images: ensemble false-positive reduction using a convolutional neural network technique. Med. Phys. **43**(6Part1), 2821–2827 (2016)
14. Bi, L., et al.: Automatic detection and classification of regions of FDG uptake in whole-body PET-CT lymphoma studies. Comput. Med. Imaging Graph. **60**, 3–10 (2017)
15. Zhao, X., et al.: Tumor co-segmentation in PET/CT using multi-modality fully convolutional neural network. Phys. Med. Biol. **64**(1), 015011 (2018)
16. Bradshaw, T., et al.: Deep learning for classification of benign and malignant bone lesions in [F-18] NaF PET/CT images. J. Nucl. Med. **59**(supplement 1), 327–327 (2018)
17. Zhang, W., et al.: Deep convolutional neural networks for multi-modality isointense infant brain image segmentation. NeuroImage **108**, 214–224 (2015)
18. Moe, Y.M., et al.: Deep learning for automatic tumour segmentation in PET/CT images of patients with head and neck cancers. arXiv preprint arXiv:1908.00841 (2019)

19. Milletari, F., Navab, N., Ahmadi, S.-A.: V-net: fully convolutional neural networks for volumetric medical image segmentation. In: 2016 Fourth International Conference on 3D Vision (3DV). IEEE (2016)
20. Goodfellow, I., et al.: Generative adversarial nets. In: Advances in Neural Information Processing Systems (2014)
21. Xue, Y., et al.: SegAN: adversarial network with multi-scale l 1 loss for medical image segmentation. Neuroinformatics 16(3–4), 383–392 (2018)
22. Hung, W.-C., et al.: Adversarial learning for semi-supervised semantic segmentation. arXiv preprint arXiv:1802.07934 (2018)
23. Kim, B., Ye, J.C.: Mumford-Shah loss functional for image segmentation with deep learning. IEEE Trans. Image Process. 29, 1856–1866 (2019)
24. Andrearczyk, V., et al.: Automatic head and neck tumor segmentation in PET/CT. In: MICCAI 2020 (2020)
25. Luc, P., et al.: Semantic segmentation using adversarial networks. arXiv preprint arXiv: 1611.08408 (2016)
26. Kim, B., Ye, J.C.: Cycle-consistent adversarial network with polyphase U-Nets for liver lesion segmentation (2018)
27. Ye, J.C., Han, Y., Cha, E.: Deep convolutional framelets: a general deep learning framework for inverse problems. SIAM J. Imaging Sci. 11(2), 991–1048 (2018)
28. Kingma, D.P., Ba, J.: Adam: a method for stochastic optimization. arXiv preprint arXiv: 1412.6980 (2014)
29. Gibson, E., et al.: NiftyNet: a deep-learning platform for medical imaging. Comput. Methods Programs Biomed. 158, 113–122 (2018)
30. Weisman, A.J., et al.: Convolutional neural networks for automated PET/CT detection of diseased lymph node burden in patients with lymphoma. Radiol. Artif. Intell. 2(5), e200016 (2020)
31. Myronenko, A.: 3D MRI brain tumor segmentation using autoencoder regularization. In: Crimi, A., Bakas, S., Kuijf, H., Keyvan, F., Reyes, M., van Walsum, T. (eds.) BrainLes 2018. LNCS, vol. 11384, pp. 311–320. Springer, Cham (2019). https://doi.org/10.1007/978-3-030-11726-9_28
32. Capobianco, N., et al.: Deep learning FDG uptake classification enables total metabolic tumor volume estimation in diffuse large B-cell lymphoma. J. Nucl. Med. (2020). p. jnumed. 120.242412
33. Zhu, J.-Y., et al.: Unpaired image-to-image translation using cycle-consistent adversarial networks. In: Proceedings of the IEEE International Conference on Computer Vision (2017)

Author Index

Printed in the United States
By Bookmasters